SpringerWienNewYork

Studies in Space Policy
Volume 2

Edited by the European Space Policy Institute
Director: Kai-Uwe Schrogl

Kai-Uwe Schrogl, Charlotte Mathieu,
Agnieszka Lukaszczyk (eds.)

Threats, Risks and Sustainability – Answers by Space

SpringerWienNewYork

Kai-Uwe Schrogl
Charlotte Mathieu
Agnieszka Lukaszczyk

SpringerWienNewYork is a part of
Springer Science + Business Media
springer.at

Typesetting: Thomson Press (India) Ltd., Chennai

Cover: Eumetsat (Metop-A. AVHRR RGB Composite. 9 September 2008. Hurricane Ike over Cuba)

Printed on acid-free and chlorine-free bleached paper

SPIN: 12527823

With 61 Figures

Library of Congress Control Number: 2009924861

ISSN 1868-5307
ISBN 978-3-211-87449-3 SpringerWienNewYork

Table of contents

CHAPTER 5 Energy

CHAPTER 6 Mobility

CHAPTER 7 Security

CHAPTER 8 Conclusions and Recommendations

CHAPTER 1

SETTING THE STAGE

1.1 Space and sustainability

Kai-Uwe Schrogl, Charlotte Mathieu & Agnieszka Lukaszczyk

1.1.1. Sustainability and space

Achieving sustainability is one of the great policy objectives of our time. It is a global concept and it comprises a huge set of policy areas. Its main idea is to maintain the longevity of the global ecosystem safeguarding humanity's further development or even survival. Achieving sustainability does not only rest on political will. But it also needs instruments and this book is about a specific instrument, which would provide a sound basis for decision-makers to find the right solutions: the use of space, in more concrete terms the use of application satellites. The aim of the contributors to his volume is to demonstrate the benefits that can be derived from the use of satellite applications in coping with the threats and risks today's global society is facing and how space applications can provide efficient tools for dealing with the policy challenges to achieve sustainability.

The potential of space applications – telecommunications, navigation and Earth observation – either used separately or in integrated ways, has become visible on a global scale only during the past decade. Still in 1992, when the Earth Summit in Rio de Janeiro[1] gave the signal for tackling the global environmental problems, space or satellites were not prioritised in the conference itself or in the final documents. Since then, this has considerably changed. Stepwise, satellites have not only developed into operational tools but they have also been able to market their capabilities vis-à-vis the decision-makers, who do not always embrace high technology. An important milestone has been the Third United Nations Conference on the Exploration and Peaceful Uses of Outer Space (UNISPACE III) of 1999.[2] This conference provided a comprehensive outline and outlook on the potential of satellite applications for development and sustainability. It also invigorated a close coordination of the institutions inside the UN system for the use of satellite technologies in order to achieve their respective goals.

The years since then have been filled with a systematic enlargement of the role space plays in ensuring sustainability. This is true for national approaches, where in particular India is an impressive frontrunner in methodically using indigenous space activities in fields such as environmental monitoring, resource management or tele-education and tele-health. It is also true for the global level, where more and more specialised agencies of the United Nations, like the Food and Agriculture Organization (FAO) or the World Health Organization (WHO) and of course the World Meteorological Organization (WMO) now use these technologies much more broadly. In addition to that, the role satellites play in monitoring the implementation of international agreements has made them already indispensable – a development which was not foreseen when the UN Framework Convention on Climate Change was negotiated in 1992.

Following this development and building on these achievements,[3] it seems appropriate to review the status and the prospects for space technologies[4] in further supporting sustainable development and in dealing with the threats and risks humankind is facing in particular due to climate change. For this purpose, this volume applies a novel approach in showing, in a systematic way, the contributions of space technologies but also the policy requirements for fulfilling the growing expectations. It divides sustainability into six issue areas: knowledge, environment, resources, energy, mobility, and security. This division regards sustainability as an intersection of these policy and issue areas. Locating space applications in these issue areas then provides for a clear picture of their contribution to risks, threats, and sustainability. It also has the advantage of identifying the relevant policy mechanisms for dealing with the challenges space applications face in these issue areas. The volume therefore does not stop at describing the benefits provided by satellites but also indicates the policies needed for ensuring that satellites are used to their full potential. The division into these six issue areas sets the basis for properly addressing the right decision-making forums.

1.1.2. The European Space Policy Institute's (ESPI) conference "Threats, Risks and Sustainability – Answers by Space"

This volume is based on a conference organised by the ESPI at its premises on 10–11 December 2007 in Vienna, Austria. The task of ESPI as an independent European think tank is to facilitate the decision-making process and to increase awareness on space technologies. ESPI also pays great attention to opening the narrower field of space policy to other policy and issue areas. The conference on "Threats, Risks and Sustainability – Answers by Space" was one distinct activity in this context. This conference tried to shed a new light on threats, risks and sustainability by combining approaches from various disciplines. It analyzed what could be the contribution of space tools to predict, manage, and mitigate those threats. It aimed at demonstrating that space is now pervasive and a tool in solving today's problems. More than that, it has become a visible tool in shaping society and the way we think.[5] The conference introduced various methods and ideas addressed to national and European policy makers. In the six different issue-areas, specialists in the broader field of knowledge, environment, resources, energy, mobility, and security nourished the dialogue with space experts to examine and illustrate in detail how space contributes to sustainability on Earth and how this can be made even more efficient and focused.

Interaction between space experts and analysts of broader issue areas was therefore an important aim of this conference. Thus, this volume does not simply comprise the presented papers but has also provided the speakers with the opportunity to reflect on this exchange in their final contributions to this book. In each of the six sections, an overview article is followed by two articles focusing on space. All sections are introduced by short summaries and as a final conclusion, Part 8 explains the challenges ahead and suggests tasks to be addressed by decision-makers. As an introduction to this volume, an article by the French academic Jacques Blamont sets the scene, with a view on the current 'century of threats,' as one book of this notable thinker is entitled.[6]

The ESPI conference took place at the same time as the UN Climate Change Conference in Bali, which adopted the Bali Roadmap for a future international agreement on climate change.[7] Bali showed how indispensable space technologies are today for addressing the challenges of climate change. Our book intends to extend the impression of space technology's benefits for the environment as well as knowledge, resources, energy, mobility, and security. The efficient and systematic use of satellite telecommunications, navigation and Earth observation, separately or integrated, are keys to sustainable development. They are ready to support the prevention, monitoring, anticipation and management of threats, risks, and occurring disasters. They are available for detecting resources, managing food production, identifying epidemics, or supporting the establishment and running of plants for hydro or solar power as well as power grids. And finally, space helps securing peace through supporting confidence-building measures and provides humankind with knowledge about our universe and the place of Earth in the solar system. Maybe it is the view of Earth, for the first time provided by satellites and later communicated by astronauts, which best reminds us that we are one world and have to act accordingly.

---

[1] UN Conference on Environment and Development (UNCED), 3–14 June 1992.

[2] See in particular the document: Space Benefits for Humanity in the Twenty-First Century: A Compilation of Contributed Papers for the UNISPACE III Background Papers. UN Doc. A/CONF.184/BP13, 1999. For an analysis, see: Liebig, Volker, and Kai-Uwe Schrogl. Space Applications and Policies for the New Century: The Impact of UNISPACE III 1999. Frankfurt/Main: Peter Lang, 2000.

[3] Annual overviews are provided by the series "Highlights in Space" prepared by the UN Office for Outer Space Affairs (UNOOSA) in cooperation with the International Astronautical Federation (IAF), the Committee of Space Research (COSPAR) and the International Institute of Space Law (IISL). For the latest edition covering the year 2007, see: UN Doc. ST/SPACE/37, 2008. See also: UNOOSA. Space Solutions for the World's Problems: How the UN Family Uses Space Technology to Achieve Developmental Goals. UN Doc. V.06-57190, 2006; and Peter, Nicolas. "Space Policies, Issues and Trends 2007/2008." ESPI Report 15. Vienna: European Space Policy Institute, 2008.
[4] See also: Feuerbacher, Berndt and Heinz, Stoewer, eds. Utilization of Space: Today and Tomorrow. Berlin: Springer, 2006; and: Organization for Economic Co-operation and Development, ed. Space 2030: Exploring the Future of Space Applications. Paris: OECD, 2004.
[5] See for example: Dick, Steven J. and Launius, Roger D. eds. Societal Impact of Spaceflight. NASA SP-2007-4801. Washington, D.C.: NASA History Office, 2007; or: Pompidou, Alain, ed. The Ethics of Space Policy. Paris: UNESCO, 2000.
[6] Blamont, Jacques. Introduction au Siècle des Ménaces. Paris: Odile Jakob, 2004.
[7] Report of the Conference of the Parties on its Thirteenth Session, held in Bali from 3 to 15 December 2007, UN Doc. FCCC/CP/2007/6.

1.2 The synergy of global threats[8]

Jacques Blamont

1.2.1. The basic facts: population growth[9]

The world's population that stands today at 6.6 billion will reach 9.2 billion by the middle of the 21st century, and then continue to increase by 30 million every year. This development is characterised by the contrast between developing countries where growth is strong, and developed countries (with the exception of the U.S.) where the population is stagnating or decreasing. By 2050, 80% of the people will live in developing countries (5.2 bn in Asia, 2 bn in Africa, Europe will be down to 0.66 from 0.73).

We will call *gradients* zones where population density varies rapidly from one region to another. Today's situation is remarkable due to the quasi-instantaneous birth of gradients. The major trend is fast urbanisation, with the formation of megapoles containing 10–30 million inhabitants. Each week, 1.3 million people leave the countryside for a city. We already have 428 cities with more than one million inhabitants. Of the 60 agglomerations above five million, 48 are situated in developing countries. It is estimated that 80% of the world's population will be urbanised in 2050–2060. At that time, 3.5–4 billion people of 9 billion shall be "housed" in shanty towns.

Let us quote an example of importance for Europe: In 2050, the Maghreb will count 105 million people of which more than half will dispose of less than two euros per day. The countries of North Africa, from Morocco to Suez, line up in the range of 4–8 euros per day and head, which has not changed in 50 years, and their resources will not increase on the northern shore of the Mediterranean, faced with the Maghreb's explosive demography. The population of Europe will be characterised by old age and low birth rates: Germany, Italy, and will have lost 10% of their present population while still enjoying incomes of 70–80 euros per day and head. The North–South gradient will not be able to generate socio-cultural consequences. A recent CIA report estimates the percentage of Muslims in the European population at 30% by around 2030. Europe will by then be different from what it is today.

The same situation prevails in Southeast Asia, where Indonesia is creating demographic pressure on its neighbours Malaysia and Singapore from which hundreds of thousands of immigrants are currently being expelled. Another example is Palestine: Palestinians have an income of 2.50 euros per day (3.50 euros with international help) and Israelis 45 euros, and they are separated by just a line. Despite any goodwill and nice words, this situation is not negotiable.

1.2.2. The limits of productivism

1.2.2.1. Agriculture

Today, more than 800 million people living in developing countries receive insufficient food. Among the three billion people subsisting on agriculture in the world, one-fifth does not eat properly, as regards neither quantity nor quality. Bearing this fact in mind that while mankind counts six billion souls, we are led to wonder if it can produce enough goods and services for nine billion in 2050, and how will agriculture meet these needs.

After the Second World War, the fear of a major food crisis was proven unfounded by the advent of the Green Revolution, which gave self-sufficiency to Asia. The Green Revolution may be defined as:

- A combination of production techniques for agriculture including irrigation and fertilizers;
- A combination of social support policies in favour of the peasants.

The Green Revolution succeeded in the irrigated zones of humid tropical regions, but only partially in regions of pluvial agriculture, and not at all in dry regions where poor farmers did not have the possibility of investing enough to overcome climatic vagaries. It has helped familial farming in all of Asia, but much less in South America where its techniques have yielded profits for large estates. In Africa, it has not reached the average producer.

Whatever the performance achieved in Asia, it is legitimate to note that, outside of any quantitative considerations, the Green Revolution has caused the exile of rural populations, destroyed soils, waters, and social roots. These dislocations have bred the explosion of births, the urbanisation, the loss of social relations, thus unleashing a new cycle of impoverishment and starvation. Land degradation has reached an alarming level due to land shortage, inappropriate land use, severe economic pressure on farmers, poverty and population growth according to UN reports. Globally, 4700 million hectares are affected. In South Asia, 43% of the region's agricultural land suffered some form of degradation: of 140 Mha, 31 Mha suffered strongly and 63 Mha moderately. In Iran, 94% of the agricultural terrains are degraded, 75% in Bangladesh, 61% in Pakistan, 44% in Sri Lanka, 33% in Afghanistan, and 25% in India.

In India, of 329 Mha, 143 are sown and 75 irrigated, compared to 5 before the Green Revolution. 107 are subject to various kinds of degradation (in 1994), water erosion being the major factor (57 Mha), followed by wind erosion (10.5 Mha). A new phenomenon has recently appeared and is increasing steadily: the salinification and acidification of soils. This is well measured by satellite remote sensing, hitting now 20 Mha up from 7 Mha in 1987 and 10 Mha in 1993. Crop yield is diminishing, and India had to import rice for the first time in 2005.

What is needed to feed 9 billion people? A study by P. Collomb[10] for the FAO shows that Asia and South America need to double and Africa quintuple their production by 2050. The totality of resources seems to be insufficient with available techniques. Furthermore, with this new distribution of productive zones, international specialisation should be harmonised so as to enable potential importer nations to also export goods on international markets in order to be able to buy their food.

How can we repeat the success of the Green Revolution in a way that is compatible with the environment, economically viable and based on more social equity? This objective has been called the "Double-Green Revolution" and would consist of reducing the chemical damaging of the productive ecosystems and simultaneously granting access rights to land, resources and social services to poor farmers. This hazy concept belongs to the realm of dreams.

1.2.2.2. Water

In the year 2000, 8,000 km^3 of water per year are needed to quench the thirst of 6 billion people, including 5,000 of rain over 1.34×10^9 ha (pluvial agriculture) and 3,000 provided by rivers and water tables over 0.27×10^9 ha (irrigated agriculture).

In the year 2050, 10,000 km^3, maybe 12,000 km^3 if meat consumption increases, have to be added to 4,000 to reduce today's deficit, which makes 16,000 km^3, or 14% of rainfall. Thus, globally, humankind should not suffer from a water shortage in the near future. However, in contrast to this optimistic statement, the consequences of the demographic expansion are not related with the quantity of water, but its availability in the right places when the population grows. It is again a problem of gradient. Today, an unprecedented number of people, that is 1.1 bn, have no access to drinking water, 2.6 bn do not dispose of purification or decontamination facilities, and 3 bn have no tap water at home.

Two-thirds of the water is used for agriculture. The first effort should be to increase efficiency, but the best saving we can expect in this domain for 2050 does not exceed 20%. Could the solution be to increase areas of irrigated agriculture? With the present increase of 1.34 Mha per year, the present 264 Mha will become only 331 Mha by 2050. It would be necessary to multiply the present growth rate by 10 to satisfy demand and it is obviously out of reach. The only possibility is a large increase in pluvial agriculture, which occupies now 1.34 Bha and should grow by 1 Bha by 2050; adequate surfaces would be available in sub-Saharan Africa, South America and Russia: these areas will have to feed the world. In this scenario, the necessary major clearing of zones covered presently by natural vegetation would strike an irreversible blow to biological diversity. Already 15 Mha of forest disappear every year in Brazil, while 12–16 Mha of arable soil are lost by urbanisation, erosion, and salinification.

Many countries in Asia, the Middle East and Africa will not be able to feed themselves and will have to import food from South America. There is no solution adapted to the magnitude of the problem, including the development and use of genetically modified organisms.

Tapping the underground reserves has already started on a large scale. The aquifers below the Northern China plain will be drained within 30 years. India will have used up all its regulating stocks in less than 20 years, and what will remain is only the yearly flux infiltrated during the preceding humid season, highly variable from one year to the other. Overexploitation, that is excess of consumption above reloading, is the rule in all countries of the Middle East, in Turkey, and in Mexico. In India and Pakistan, 60% of the population depend on ground water, 25% in China, and 15% in Iran.

In order to use seawater, it is necessary to desalinise it. The cost of this operation depends on the price of energy: 0.8 dollar per m^3 for an oil barrel at 60 dollar (acceptable for drinking water, but 50 times too expensive for agriculture – the main market). In order to use desalinisation to satisfy the need for food of 3 billion more people estimated at 4,000 km^3 per year by 2050, we would have to spend per year the equivalent in energy of three times the quantity of fossil hydrocarbides presently used at the rate of 2.5 kW/m^3, i.e., 7% of the world's GNP.

Access to water is complicated for nations that share basins. The 263 international basins are crossed by the political boundaries of two countries or more. For example, the Nile basin is shared by 10 countries. If the past history shows that the quarrels related to water have always been solved peacefully, even between enemies, today conflicts could be fed by two new factors, the growth of the population and climatic change. The needs of India and China for domestic use of water will double by 2030, with a similar trend for industrial needs. During the same period, the hydric machine will accelerate locally, precipitation will decrease here and increase there, the succession of climatic zones will move towards the poles, and the frequency of extreme meteorological events such as droughts or floods could increase.

1.2.2.3. Fossil fuels

The world's demand for energy increases yearly by 2%: 1.5% for developed nations and 5% for emerging nations. Consumption will grow from the present 9 Gtoe[11] to 18 (15–20 seems a reasonable range) in the year 2050. At present, a U.S. citizen receives 8 toe/year, a European 4, and an Indian 0.6. The figure of 18 Gtoe is obtained assuming three to four toe/year/head for developed countries (a reduction ...) and 1/5 to 2 toe for the 8 billion people of the poor nations.

Since the proportion of production based on fossil fuels is 90%, the consumption in fossil fuels will also increase yearly by 2%. The ratio of proven reserves to the consumption, i.e. the number of years during which present growth can be sustained, is 40 years for oil, 50 for gas, and 100 for coal. By 2050, fossil fuels will not provide more than 12 Gtoe. The other sources (nuclear, renewables and savings) will hardly produce more than 2 Gtoe. In order to provide the missing 4 Gtoe, 2,500 nuclear power plants would have to be added to the existing 450: one would have to open one every 10 days! The decrease in resources will increase the price of fossil fuels with a major impact on developing countries.

1.2.2.4. Biodiversity

Since the appearance of molecular oxygen in the atmosphere 600 million years ago, five major crises, called mass extinctions, have marked the evolution of living species. Outside of crises, species appear and disappear at the ratio of one species every four years (natural extinction). Presently, one and a half million species are known, of a possible total of a few million. Under man's influence, the yearly extinction rate has suddenly increased to 10,000 per year, at least 40,000 times the natural extinction rate. The idea of a "6th extinction" has been gaining ground.

In 2000, more than 3,500 species of vertebrates and invertebrates living in continental waters were classified as being at high risk of extinction, and more than 20% of the fresh water species were threatened. For terrestrial animal species, their present rate of extinction is related to the diminishing tropical forest where the majority of them live. Man is destroying the forests at the rate of 1% of its surface per year. Around the year 2050, when forests will have been reduced to 10% of their original extension, 50% of the species sheltered there will have disappeared. The EPUN announced in 2001 that 25% of mammals (1130 species) and 12% of the birds are threatened.

The state of an ecosystem fluctuates around average values. The cruise mode can suddenly be interrupted by the growth of one parameter and be replaced by a new regime. Such crises are called bifurcations or catastrophes, and are usually irreversible, just as the initial state can only be reinstated for a different value of the critical parameter. The proper management of our relations with ecosystems requires the identification of the critical variables. Among them, biodiversity is an essential requirement of aptitude to regeneration after perturbation and an insurance against risks. And we see that biodiversity is at risk.

1.2.3. The new players

The trends towards an unmanageable relationship between the developed world and its environment, perceptible since the year 1970, have been amplified by the arrival of new players converted to the doctrine and practice of productivism.

1.2.3.1. China

China has adopted productivism to hoist its 1.3 billion citizens to the GNP of the industrialised nations. Its growth rate, fed by exports and construction, has surpassed 10% per year for the last 10 years (10.7% in 2006). With 700,000 engineers and scientists, China claims to be the second scientific and technological nation after the U.S. and intends to become the first before 2020.

China's steel production now exceeds that of Germany, Japan and the United States combined (38% of the world's total). Automobile production is a symbol of industrial power. Presently, China counts 25 cars to 1,000 inhabitants (800/1,000 in the USA). But China, with seven million cars sold in 2006 (at a production capacity of 8 m), has reached position number three worldwide, ahead of Germany. Its objective is 250 million cars. The world is moving from 600 million cars towards one billion. China has fixed the year 2020 for launching a large commercial airplane, and after having placed two astronauts in orbit, it plans to conquer the Moon.

However, the natural resources required for achieving this development aimed for by the authorities are lacking, except for coal which provides 70% of the energy of the country. With 118 billion tonnes of reserves (13% of the world's), the country can continue for a half century at the present consumption rate of 2 billion tonnes a year. Since coal is the fuel that emits the largest quantity of CO_2 per toe, China will surpass the U.S. in 2009 as the largest CO_2 producer in the world. In addition, its energetic consumption will double by 2020. The country is covered with power plants with an average capacity of 25–600 MW and opens a new one every 10 days. They all operate with immature, inefficient, and costly processes. The energy problem of China is only just starting to appear.

Its enormous imports of oil have already become a political factor in the Middle East and in Africa.

Apart from the loss of arable soil, air pollution, and the destruction of the environment, there is also the question of water scarcity. Water usage has quintupled since 1949. The resources per head are below one-third of the global average: China has 7% of the world's water resources and 20% of its population, a situation complicated by a severe imbalance, with four-fifths of the water supply in the south. In the plains of the north where 43% of the population live, people must make do with 14% of the water for producing half the wheat and one-third of the corn. As a leading producer of wheat ahead of the U.S., China is due to becoming an importer of this cereal. China already imports soy, because the water table is diminishing due to over-pumping. Four hundred of its six hundred cities do not have the water they need for growth. The price of access to water is estimated to become prohibitive by 2030. Presently, agriculture absorbs 70% of the water supply. A project of pharaonic dimension has started the yearly transfer of 45 trillion litres of water from the Yang Tse Kiang to the north, but this resource may have to be utilised in the cities to cover the vital needs of the population and industry. "We have a water shortage, but we have to develop", said a senior engineer at a Chinese city's water conservation bureau to a New York Times reporter. "And development is going to be put first".

In 2025, China will land a Taikonaute on the Moon's surface. This will help the 350 million Chinese who will have no drinking water.

1.2.3.2. India

Today, India has 1.1 billion inhabitants with a birth rate of 3.2 children per woman; their number increases by 1.2% per year and will surpass China's in 2030. Four hundred million Indians live on less than one euro per day, and just as many with less than two euros. Two-thirds of the active population are farmers in 680,000 villages. Mixed with those masses, there is a class whose standard of living is similar to that of the industrialised countries and whose number is estimated at 50–100 million. India is a divided country.

After 1991, a growth process began based on the adoption of the productivism model. Annual growth reached 8.2% in the last 3 years, and should be maintained at 9% for the next 10 years, even if the growth rate of agriculture stagnates around 3%, a figure which confirms the difficulties created by the Green Revolution.

The production of energy should follow the overall growth rate, at the 8% level. Coal accounts for 55% of the consumption of fossil fuels. The insufficient production of 355 Mtonnes will augment by 20 M per year and reach 700 million by 2015. At this rate, the 93 billion tonnes of reserves will be exhausted in 100 years.

The annual growth of the automobile market is 20% per year, with only one million cars sold in 2005. Since the proportion is seven cars/1,000 inhabitants, the potential market is enormous. Foreign companies like General Motors have announced investments of four billion U.S. dollars. Forty billion dollars are needed to increase the cash flow of the automobile sector to 145 billion dollars, the government's 10-year target. This money will be found.

At the beginning of the year 1990, the country had only eight million telephone connections. Now, 140 million Indians have a mobile telephone (450 million in China) and the figure will reach 500 million by 2010, since seven million people buy one every month. It is in software that India has achieved extraordinary progress, capturing 50% of the world's software subcontracting, promoting its services in transport, the tele-managing of enterprises, and health. In 2006, 35% of exports were reported by 0.06% of the population, the software specialists.

Indian growth does not profit all Indians since, far from generating an industrial revolution, it is based on added value services not requiring heavy manpower. The software domain employs only 1.6 million individuals, as compared to the eight million jobs that should be created every year. Here like everywhere else, the expansion of modern activities in cities attracts farmers who abandon their homes to go to the towns. This development is hard to accept in the countryside: in 2005, 10,000 farmers committed suicide.

As in the former Soviet Union and like in China today, the Indian conversion to productivism is accompanied by a disregard for the environment. As in China, development rests on coal. In the countryside, Chinese diesel generators have become a necessity for the poor, cheaper than coal despite being dirtier and less efficient. Another factor in India as in all of Southeast Asia, air conditioning consumes vast quantities of energy. In November 2006, the Indian Space Agency announced that work would soon begin for a man-in-space national programme.

1.2.3.3. Conclusion

India and China have a total population of 2.5 billion, including 1.5 billion people who live on less than two euros per day, and follow a productivist course of development. The impact of their economies on the rest of the world will grow:

- Their objectives cannot be reached without a fast increase in exports,
- Their resources are limited: in energy, their lack of oil is forcing them to support the oil-producing countries and is driving up the price of fossil fuels; their lack of water can only create conflicts,

- Thousands of their enterprises are active outside their national markets, their multinationals invest heavily in Western countries,
- Their cheap labour accelerates outsourcing and creates unemployment in developed countries.

1.2.4. Changes to the way of life

1.2.4.1. Old age

As soon as in 2025, based on the progress of medicine, there will be 820 million people older than 65 in the world instead of 470 million today. Old age appears to be one of the epidemics of the 21st century.

The Ancient Greeks used to consider the age of 80 as the ultimate limit of human life. Our vision and our practice are based on the idea that this age can be reached and exceeded by everybody. In the year 1800, the first proven centenary died; they will be 150,000 in France by 2050.

Half of the health expenses are incurred in the last years of life. Among the organs that become fragile due to age is the brain. In the U.S., 4 million patients suffer from Alzheimer's disease, and this figure will climb to 14 million by 2050. Already, the annual cost is above 100 billion dollars. France will have to support 1.2 million senile persons older than 85 in 2050: who will pay?

1.2.4.2. Epidemics

Mankind has become more exposed to the spread of old and new diseases since people travel more, faster, and to more remote areas. Around 2020, 1.5 billion tourists, double the present figure, will travel across the planet: microbes are no longer confined to faraway ecospheres or rare reservoir species, but they circle the Earth in 24 h. Our times are witnessing the outbreak of unknown bacteria or viruses, their migration to new regions, their dissemination through technology, and their jump from animal to man. HIV/AIDS is the symbol of modern epidemics caused by our customs and globalised by transport. Since 1973, thirty new infectious diseases have appeared, while major traditional endemics have been coming back such as malaria, tuberculosis, yellow fever, and cholera. Koch's bacillus infected one-third of the world's population, killing three million people in 1995. The WHO admitted in 1992 that they had no strategy against malaria, contracted each year by 500 million victims. Confronted with the increasing resistance of germs, we should listen to the Harvard epidemiologists who in 2000 published the warning: "We are on the verge of a world crisis as far as infectious diseases are concerned".

1.2.4.3. Climatic change

The increase in CO_2 emissions by human action, identical for an obvious reason to the increase in the production of energy, is estimated presently at 2.1% yearly. CO_2 produces a greenhouse effect which adds to the natural greenhouse. The scientific community is today quasi unanimous in estimating that this trend will induce climatic changes. The effect will not be benign because a large quantity of energy will be injected into the atmosphere, and its most important characteristic is its speed, with a time frame too short for the ecosystems well-adapted to the present climate to adjust.

Without describing the consequences zone by zone, let us mention that 6–10% of the area of cultivable soils will be lost in Europe in the coming century. Although weaker in the Mediterranean area than at high latitudes, global warming will still have a major socio-economic impact that will be strongly negative. Six to fourteen million Europeans might live in regions dominated by hydric stress by 2020 in contrast to their present satisfactory water supply. Developing countries should suffer most.

The effects of climatic change will have an impact on health. Warming expands the territory of the vectors of infectious diseases. For instance, mosquitoes thrive in warm atmospheres, because the rising night and winter temperatures diminish their mortality threshold.

Thus, the climatic impact of the greenhouse effect created by the energy binge of mankind appears as deeply synergistic: it will affect all ways of production, the way of life, the satisfaction of basic needs, health and the environment. Again, developing countries will suffer more than developed ones: an example was given in 2004 when thousands of people died in Haïti during hurricanes, while Florida's population was spared under the same conditions. The differences between the North and South will increase, generating political instability and conflicts.

What can we do about climatic change? The CO_2 emissions of France were less than 1.5% of the global ones in 2000, the EU's (25 countries) were 14%, targeted to become 7% around 2050 since growth will be happening elsewhere (U.S., China, India). The major actors cannot abandon fossil fuels, and especially coal. The Kyoto Protocol, which is supposed to limit emissions, is not being implemented, and, if it were, it would have no effect. Climatic warming, now a fad of the media and of politicians, will not be stopped and might reach 2–4°C in the coming century, with painful consequences, as throughout history, the human enterprise has rested on the stability of the climate. However, this is only one factor among many other social diseases generated on a global scale by productivism, that is, by the mania of development.

1.2.4.4. Transportation

Ninety-eight percent of all transport relies on oil, with a global increase of 2% per year. The volume of consumption by 2050 will be multiplied by three for freight and two for passengers. With the evolution of the price and availability of oil, how can we save our noblest achievement: the automobile?

There has been much hubbub about bio-fuels, ethanol and diester supposed to replace fossil fuels. Ethanol provides 25% of the energy needed to achieve that, and diester 93%. It has been said that the production of one ton of ethanol requires one ton of oil.

The cultivation of vegetables demands large areas: the yield is 0.75 toe/ha/year for all options excluding sugar cane, which stands at 4.5 toe/ha/year. Sugar cane in tropical zones is therefore the only interesting option for fuel production. Brazil already produces 15 Mtoe of ethanol, and hopes to obtain 100 Mtoe by 2030 without touching the Amazonian forest. The corn craze in the U.S. is hard to understand.

The world has 1.4 billion ha of arable lands and uses 3.5 Gtoe for transportation. It could therefore not provide more than 1.4 Gtoe at maximum: if we were to stop eating, we could feed 40% of our cars with bio-fuels! Let us conclude that 15–20% of our demand for transportation could be covered by bio-fuels. Today. And with a 2% yearly increase . . . ?

1.2.4.5. Disasters

The number of major natural disasters such as floods, storms and droughts has quadrupled since 1960 and economic losses have been multiplied by eight, according to insurance companies, who expect this trend to continue in the coming years.

The developing countries are most stricken by natural disasters. From 1970 to 1997, 75% of the major catastrophes happened in the region of Asia Pacific. Of 1.4 million victims in the last 20 years, 98% came from poor countries. For each casualty, 2,000–3,000 persons are affected. Developing countries are vulnerable to climatic conditions because of their important habitat in coastal zones of low altitude. By 2025, more than half of the population of South East Asia will live in risk zones.

With the globalisation of the economy, the repercussions of a disaster propagate far from the epicentre. The earthquake that hit the city of Chichi in Taiwan on 21 September 1999 provoked a worldwide scarcity of memory chips for weeks, which would have become months if the epicentre had been only 60 km away from the real location.

It is a very dangerous error that the settlement of land has expanded since the beginning of the 20th century without paying due attention to risks. Forty towns out of fifty whose growth rates are the highest are situated in zones prone to earthquakes.

1.2.5. Synergy of threats

In 1988, the concept of sustainable development was invented, the last avatar of productivism which pretends to promote access for all people to the standard of living of the industrialised nations while protecting the available resources. However, this concept encounters a fundamental limitation which it does not factor in properly: the availability of resources. Their exploitation at the present rate is not sustainable. To raise the people who live on two euros per day, that is, half of mankind, to the level of 50 euros is thus an absurdity. China's catching up alone would roughly double the world's consumption rate. Oil consumption would increase by 106%, for instance, and world metal consumption by 94%. If India as well as China were to catch up, world consumption rates would triple. If the whole developing world were suddenly to catch up, world rates would increase eleven-fold. It would be as if the world population ballooned to 72 billion people (retaining present consumption rates).

What is going to happen?

- The "50-euros-per-day-nations" will fight to keep their advantages. President Bush (first) declared in 1992: "The standard of life of the United States is not negotiable".
- The "one-euro-per-day-populations" will feel more and more frustrated with their access through TV to world which lives at 50 euros per day.
- The unabated exploitation of natural resources will accelerate and deepen this gap even further.

The situation is not sustainable at the six billion people level. What will it be at nine?

We have seen that all trends will deepen the gap between the rich and the poor. Now, if technological progress has not brought happiness to mankind, it does level out the conflict potentials between the world's actors. A latent instability lies heavy on the international order, generated by the light-headed development of nuclear, chemical and biological weapons called massive destruction armaments (MDA).

A double threat follows: the first will comes from the developing states who feel attacked by the industrialised states in the fight for natural resources. In the domain of nuisance, the technological progress favours the smallest: the U.S. does not become stronger when they add 10 nuclear bombs to their arsenal of thousands, but the acquisition by North Korea or Iran of 10 bombs worries the world. Technological progress therefore leads to the unavoidable proliferation of MDAs.

The second threat comes not from states, but from the unorganised crowds of people earning one euro per day, and is also an effect of modern technology: the use of networks similar to the Internet. Internet is a model: Al Qaeda officially boasts to be a virtual state, even an electronic state, and serves as an example to many other networks, related to it or not. Activists of all sorts live on the Web like "fish in the water". From a tactical point of view, the technique of *swarming* has been used; it consists in the simultaneous attack by a large number of nodes towards a particular point, led by dispersed units converging or by multiple and diversified informational tools. The attempts at Madrid, London and the WTC are examples of swarming where the target is the virtual world. The war of the 'one-euro-per-day-people' unfolds on the media battlefield.

The Internet is also a tool for the 'one-euro-per-day-people' networks. Al Qaeda and Salafist groups have transferred a large part of their activities to the Web where they recruit personnel, collect funds, coordinate their actions, diffuse their propaganda, and lead their psychological war. Anybody can enter an Internet café and establish connection with a site that teaches how to fabricate explosives from current commercial products in a few minutes.

The danger lies in the convergence of these two threats, that is to say, if the people living on one euro per day acquire access to MDAs developed by states, such as chemical or biological weapons. The publication of the genome of viruses in the scientific press has become controversial since the fabrication of biological weapons requires only a small number of specialists and basic material.

Our ingenuity should enable us to stave off one kind of danger: it is the conjunction of threats growing from all sides against which defence will be difficult. The mega-trends we have listed are not independent. All these evolutions whose time constant is equal to that of technological progress, therefore to the 2 years of the law of Moore, do not leave time to develop new industrial processes, new medications, or new rules of the game before the crisis hits. This is to say that I do not believe that a technology approach will be able to save the world.

An idea seems to describe the situation:

- On the one hand, all the threats are horizontal: demographic disorders, the exhaustion of natural resources, expansion of activist networks.
- On the other hand, the answers are all vertical: the policies of growth control, the mastering of mining and agricultural production, the formation of elites, scientific research: all are of a national or semi-national nature.

Mathematicians know that the product of orthogonal functions is zero. It is clear that we have no solution for curbing the flow of history. Following this idea, we will adopt the principle that in the modern world, only networks can win against networks. Governments and organisations dedicated to collective security should imagine mechanisms and structures of multi-jurisdiction and cooperation resting on "participative production" similar to the new trends leading to Web 2.0 and other innovations which have been called Wikinomics.[12]

Can space help in this deteriorating situation? This question is tackled in the contributions to this book.

[8] Blamont, Jacques. Introduction au Siècle des Menaces. Paris: Odile Jacob, 2004.

[9] United Nations Department of Economic and Social Affairs. World Population Prospects: The 2006 Revision. UN Doc. ESA/P/WP.202, 2007. The medium fertility hypothesis is adopted.

[10] Collomb, Philippe. Une Voie Étroite pour la Sécurité Alimentaire d'Ici à 2050. Paris: Economica; Rome: FAO, 1999.

[11] 1 Gtoe is one billion tonnes 'oil equivalent;' 1 toe is one ton 'oil equivalent.'

[12] Tapscott, Don and Williams, Anthony D. Wikinomics: How Mass Collaboration Changes Everything. New York: Penguin Books, 2006.

CHAPTER 2

KNOWLEDGE

2.1 Overview

The context: Europe is a knowledge-based society. This aspiration can only be sustained when it rests on a strong fundament provided by basic science. Through a strong relationship between basic science and applied science, society can benefit in all relevant issue areas. Through basic science, the understanding of the Earth system can be improved and the even more far-reaching questions about the place of Earth in the universe and the emergence of life can be approached.

The contribution of space: space is an indispensable tool for attempting to answer the question about the origins of the universe, the Earth, and also life. But satellites do not only provide views into the distant past but also help to explain the basic scientific questions about the Earth system through Earth observation. This is where knowledge, building on basic science, enters the sustainability issue area: solar research or space weather research, for example, can provide indicators of environmental phenomena on Earth. Basic science, in particular basic space science like planetary research, is also the generator for a general interest in the field of science and engineering. This might translate the fascination of the youth into a much needed growing number of scientists and engineers. Basic science is also the incubator for linking developing countries with our knowledge-based societies, leading them to self-sustained development. Europe is particularly strong in the field of basic space science and has already contributed to an explosion of knowledge allowing not least for global participation.

The papers: in this section, three papers provide a broad and highly illustrative set of information on the contribution of basic space science to our knowledge society.

In "Knowledge: Understanding Our Place in the Universe", Maria Teresa Lago, professor for astrophysics at the University of Porto and member of the European Research Council, starts with reflecting on the Earth's place in the universe. She then turns to her engagement in European science policy and states that space research visibly involves frontier research, going beyond the categories of basic and applied science or technology. In her view, space research also implies the absence of disciplinary boundaries, and most naturally complies with an intrinsically risk-taking character of groundbreaking research (high-risk/high-gain) at and beyond the frontiers of knowledge.

In "Sustainability on Earth: Space Sciences Supporting Related Research", Jean-Claude Worms, the Executive Scientific Secretary of the European Science Foundations' European Space Science Committee, illustrates the fundament laid out by basic science on which space applications are built. He also points out that applications can even emerge from the unexpected results of basic research outside the original field, thus making the point for a robust freedom of academic research.

In "Knowledge, Data and Global Sustainability: The Role of Space-Based Assets", Mark Williamson, a well-known space technology consultant, provides an extended tour of the problems which can be efficiently tackled with space applications. He also reflects on disaster management and even the deflection of asteroids menacing the Earth. In addition, he points to the task of not only achieving sustainability on Earth but also to implement a sustainable use of outer space, inter alia through the mitigation of space debris.

2.2 Knowledge: understanding our place in the universe

Maria Teresa Lago

2.2.1. Introduction: knowledge and sustainability

The topics chosen for this publication – environment, security, mobility, knowledge, resources, and energy – identify some of the main priorities of our day and are simultaneously some of the foremost areas of concern. Consequently, they are often the focus of concerted attention and debate among researchers, developers, and decision-makers. It is therefore a real challenge to address "knowledge" in such a wide context.

The following article approaches the broader topic of knowledge related to sustainability in two steps. It first looks at our knowledge of the universe as the foundation of our general understanding. It then turns to the organisation of stimulating knowledge in the European context. As a member of the European Research Council (ERC), I reflect on the work of this newly established institution and point out the challenges it is facing. This part also reflects on the importance of supporting knowledge about our place in the universe.

2.2.2. Our knowledge of the universe

This investigation starts by discussing knowledge and space through an obvious approach: our place in the universe. This immediately brings to mind what may be considered one of the most momentous moments in mankind's endeavours in this context: Galileo's use of a telescope for the first time, almost 400 years ago. His subsequent discoveries constituted a real revolution in knowledge. To mention just a few fantastic examples of his achievements: the identification of craters and mountains on the moon; spots on the surface of the sun; the rings around Saturn; and the phases of Venus.

However, the greatest impact was made by the discovery of the four moons orbiting Jupiter, which were carefully registered in Galileo's diary of observations. These findings certainly constituted the indisputable proof of the non-geocentric concept of the cosmos. There is no doubt that this clearly corresponded

Fig. 1. *Satellites of Jupiter – Io, Europa, Ganimedes, and Calisto (source: NASA).*

Fig. 2. *Another breathtaking example is this unique image of an eruption on the surface of Io registered from space (source: NASA).*

to what would nowadays be considered "frontier research": the benchmark of the European Research Council (ERC).

When looking at recent images of the same four satellites of Jupiter – Io, Europa, Ganimedes, and Calisto – for example, the images shown below taken by "Galileo" (this time, the NASA spacecraft named in his honour) from a distance of approximately one million kilometres (1997), we have to wonder what Galileo's reaction would have been (Figures 1 and 2).

I propose to proceed with the following question: What are the most essential contributions of space to knowledge? As illustrated, one is probably the global and unique view of the Earth's system, which has opened new avenues in many and broad areas of science, such as

- Physical geography
- Geology
- Geophysics
- Meteorology
- Oceanography
- Climatology
- Global environmental change
- Biogeochemical cycles
- Natural resources
- Global management

among others.

Observations from space are essential for most of these fields in terms of the individual disciplines, but most of all in terms of a progressive understanding of the entire system at work, and as an incentive for significant advances in knowledge spanning several domains. Moreover, space is equally relevant for many other areas of science, ranging from fundamental physics to material sciences and biotechnology.

In the following, I will briefly focus on yet another specific field of knowledge that has been strongly influenced by space undertakings: astronomy. Space observations are essential for astronomy in different contexts. These range from peering into the most distant objects we can reach by observation to solving practical problems such as helping to identify the best sites for new ground observatories, which host the most sophisticated telescopes today. A very good example is the Paranal Observatory in the Atacama Desert in Chile, where the European Southern Observatory (ESO) has built and operates four telescope units which constitute the Very Large Telescope (VLT) (Figure 3). These telescopes, together with four auxiliary telescopes, integrate the interferometer infrastructure

Fig. 3. *VLT in Paranal Observatory in the Atacama Desert in Chile (source: European Southern Observatory).*

(VLTI). This is currently the most challenging and sophisticated astronomical infrastructure on Earth. Its scientific accomplishments have no match at the moment and are associated with an impressive number of vanguard scientific publications.

Fig. 4. *Sombrero Galaxy (VLT ANTU+FORS1) (source: European Southern Observatory).*

They result from the fact that the VLT is able to produce unique high-resolution images of very distant objects in the universe. One example of VLT high capability is the image of the Sombrero galaxy (Figure 4) one of the many galaxies in the Virgo Cluster. Despite the quality and detail of the image and besides its beauty, we must not forget that 50 million light years separate us from this structure of approximately 100,000 light years in diameter.

Space is crucial in allowing us to perform multi-wavelength observations of the electromagnetic radiation emitted by the various objects in the universe – planets, stars, galaxies, or even larger structures. To study these objects, astronomers must observe them in multi-wavelength bands, since what we observe depends very much on the "eyes" we use. Most of the radiation emitted at various wavelength bands, from gamma rays to X-rays, ultraviolet and infrared rays, is only accessible from space because the Earth's atmosphere is opaque to electromagnetic radiation over a wide range of wavelengths.

Furthermore, the atmospheric viewing affects the quality of the astronomical images obtained from the ground by degrading their resolution. Even though overriding such effects caused by atmospheric turbulence can already be done for the observations obtained from the ground, it remains highly technologically challenging today.

Therefore, knowledge about the universe and its objects and structures – either the Milky Way or external galaxies – knowledge about the processes involved in the formation and evolution of stars, and also their death and rebirth as a new generation of stars, still depends very dramatically on observations conducted from space.

2.2.3. Knowledge in the European context

This final part discusses knowledge in another context: knowledge in the context of Europe. In terms of strategy, some of the main challenges to Europe's visibility and competitiveness are called to mind:

- Declining research performance and low visibility – in many fields, United States scientists dominate in terms of high-impact publications.
- Relatively low R&D investments compared to other regions of the world. Europe's (average) investment in R&D activities is only about 1.93% of its GDP, contrasting with 2.59% invested in the United States of America and even 3.15% invested in Japan.[1]
- Europe is witnessing a growing tendency of industrial R&D moving abroad.

- There is a fragmentation of research and funding activities; there is widespread duplication of research in Europe due to the high number of research institutions,[2] leading to a lack of competitiveness and lower international visibility.
- There are limited research career opportunities in Europe, especially for young researchers – more promising career opportunities elsewhere attract many of the most promising young people to other professions or even to other regions.
- The complex administration does not help to attract and/or maintain the best researchers.

Therefore, there is a clear need for new ideas and approaches to science policy (not only in connection with space). There is a strong need for adequate and timely support for inter-disciplinary and trans-disciplinary research as well as for simpler application preparation and reporting processes, which may save researchers time and energy. Moreover, novel and unconventional approaches in terms of evaluation are also called for.

Thus, the recent decision to establish a new structure complementing other funding schemes in Europe comes very timely, such as those within the Seventh Framework Programme and those offered by the research funding agencies operating at the national level.

The European Research Council (ERC) has been created as a pan-European funding organisation reflecting the full scope of European research and scholarship. The strategy chosen by the ERC's Scientific Council for funding comprises a novel approach aiming at encouraging the highest quality research in Europe by providing support to investigator-initiated frontier research across all fields on the basis of scientific excellence. Its vision is to boost European excellence in frontier research by investing in the best researchers and ideas through competition at the European level, with excellence being the sole criterion. It strives to raise incentives towards the quality and aspirations of individual researchers, and, at the same time, to provide benchmarks and leverage towards broader structural improvements in European research.

The fundamental principle for all ERC activities is that of stimulating investigator-initiated frontier research across all fields of research, marked by ambitious objectives and addressing important challenges at the frontiers of the fields. Projects are selected on the basis of excellence and should go substantially beyond the current state of the art. For example, they should include inter-disciplinary and trans-disciplinary developments, novel or unconventional concepts, and/or approaches opening new and important, scientific, technological or scholarly horizons. This new understanding of basic research avoids distinctions between "basic" and "applied" research, between "science" and "technology", and between the "traditional" disciplines.

The instruments chosen to pursue this challenging initiative are the ERC Starting Independent Researcher Grants targeting excellent researchers who are at the stage of starting or consolidating their own independent research team or programme, and the ERC Advanced Investigator Grants earmarked for excellent innovative investigator-initiated research projects by leading advanced investigators who have already established themselves as research leaders. Applicants are welcome irrespective of their nationality, as long as they are located in, or moving to, Europe.

The author believes that space is by nature an ideal ground for a similar approach. Space research may equally involve frontier research, namely going beyond the categories of basic or applied science and technology. Space research also implies the absence of disciplinary boundaries, and most naturally complies with an intrinsically risky character of groundbreaking research, high-risk/high-gain, at and beyond the frontiers of knowledge.

I would like to conclude with a beautiful image of Earth taken from space (Apollo 17, 1972), one of my favourites: a unique planet, besides the 230 or so

Fig. 5. *The Earth from space (source: NASA).*

exoplanets discovered up to date. A fragile and complex system, without frontiers or borders (Figure 5). A clearly irreplaceable common home for life in all its diversity. The empty, cold outer space is hardly benign to life as we know it.

[1] European Commission. Key Figures 2005 on Science, Technology and Innovation. Brussels: European Commission, 2005.
[2] UNESCO. UNESCO Science Report 2005. Paris: UNESCO, 2005.

2.3 Sustainability on Earth: space sciences supporting related research

Jean-Claude Worms

2.3.1. Introduction

Riding on the wave of the infamous rocket race of World War II humankind succeeded to free itself of the Earth's gravity well and started to venture into space 50 years ago by first sending out automated hardware and animals, and then men and women. Despite various setbacks, the human presence in space, essentially in low-Earth orbit, has never really faltered during this half century. Starting with limited scientific experiments being conducted at the time of these pioneer years, space activities now span a broad spectrum of domains from astronomy, astrophysics and the observation of exoplanets, cosmology and fundamental physics, solar and heliospheric physics and the exploration of the solar system to the observation of the Earth as well as life and physical sciences in weightlessness.

Just as quantum physics pervades all of today's technology in a way essentially unnoticeable by the average citizen, space assets deliver him with equally pervasive daily services. These not only includes meteorology; the observation and monitoring of the Earth's resources, climate, weather, and natural events; but also satellite telecommunications and broadcasting or navigation. The weather forecast that is part of these daily deliveries is currently being extended to space weather forecast, i.e. to the prediction of outbursts of solar activity which is of prime importance to satellite operators and ground-based power grid controllers, not to mention its importance for future astronauts en route to Mars.

Finally, should our fate, our cosmic destiny as a species, become endangered by a "killer" asteroid, as perhaps were the dinosaurs 65 million years ago, space would remain the only tool to lift that sword of Damocles from over our heads. Thus, and in more ways than one, our society is more than ever "watching the skies" to sustain its development.

Once limited to the "hard" sciences – the only exception to this being science-fiction literature – space now also finds an echo in the humanities and social sciences: urban and land planning, risk management, disaster prevention and mitigation, security (both civilian and military), even human exploration of the solar system are all areas where the contribution from "rocket scientists" becomes clearly insufficient and must be complemented by the reflection of social, economic and political scientists, and even humanist scholars.[15]

These various facets of the use of space science assets to support the Earth's sustainability – looking "down" as well as "up" – will be addressed in this article. However, a question must be discussed first: whether the most pressing challenges of our societies, i.e. the food and water supply, the increase of urban settlement-related problems, poverty, terrorism, the increasing intensity of natural disasters, etc. can be met using space resources and space assets.

2.3.2. Removing the limits to growth?

Will humans reach – and settle on – other habitable planets? Here seems indeed to be an urge, a drive borne out of the human desire to walk beyond the horizon, to find other worlds. It looks as if "something in his nature drives (man) to the stars".[16] The obstacles that pile up before us are, however, immense: stellar distances and constraints imposed by relativity seem to forbid such trips, at least when compared to the span of a human life and unless contemplating practical suspended animation or cryogenic suspension,[17] so-called generation ships[18] or merging man and machine.[19] Suitable worlds within our reach, then, do not appear to be that frequent. Short of the possibility to effectively terraform Mars, it looks as if human settlements remain a very distant possibility in the future, and perhaps even a downright physical impossibility.

Let us assume that these obstacles would somehow be removed or overcome. There remains the solid fact that presenting space colonisation as the possibility to offer new perspectives to millions, actually billions of impoverished human beings, appears impractical and illusory. The mere scale of the enterprise would defeat any society, government or alliance of nations. If the colonisation of other planets were to happen in the future, it could initially probably only be conceived as the bold endeavour of a few daring scientists, entrepreneurs, miners, or even misfits of our societies. Such cosmic salvation will not occur for the starving millions of today; that can only come from us, here and now, down on Earth.

2.3.3. Sustainability on Earth

Space tools, on the other hand, can provide invaluable support for the task addressed at the end of the previous section. Sustainability addresses the domains of resources (e.g. water and food), energy, demography, peace and international relations, security, education, and the preservation of the existence of the Earth as a habitat for the human genus against various threats such as global warming. Multiple foreseeable threats have indeed been identified in all these domains and space science can, together with ground-based science, offer tools to help dealing with these threats through observation, prediction and management. The next logical step, however – mitigation – needs much more than space assets and requires a combination of space and ground-based efforts, public resources and, above all, political will. Although the wealth of data obtained from space to help reduce these threats is usually insufficient and must be complemented by ground-based data and assets, it does undoubtedly represent unprecedented opportunities to evaluate a situation and quantify an issue from a global perspective.

The recognition that space assets could play a leading role in the crisis management of natural hazards led the European Space Agency (ESA) and the French Space Agency (CNES) to initiating the International Charter "Space and Major Disasters" following the UNISPACE III conference held in Vienna, Austria, in July 1999, with the Canadian Space Agency (CSA) signing the Charter in 2000. Several organisations and agencies then joined the Charter between 2001 and 2007 (NOAA, ISRO, CONAE, JAXA, USGS, BNSC/DMC, and CNSA). The International Charter aims at providing a unified system of space data acquisition and delivery to those affected by natural or man-made disasters through authorised users. Each member agency has committed resources to supporting the provisions of the Charter and thus helping to mitigate the effects of disasters on human life and property.

2.3.4. Looking everywhere, from high above

Global satellite coverage of our planet enables almost instant access to a variety of images and other data; because of this, satellites have proven essential in developing a global database to describe the Earth system and its evolution. Earth observation images indeed depict our whole planet and enable large-scale phenomena to be observed with great accuracy. Satellites also remain in orbit sufficiently long to monitor gradual environmental changes. This global coverage, however, also challenges traditional notions of individual privacy and this will need to be re-evaluated by citizens and societies in the near future.[20] The ability provided by space to observe and predict the environment and, in particular, the weather and climate, has increased the importance and value of these observations and predictions for modern societies. It is estimated that up to 40% of the U.S. economy (with its GDP of 13.84 trillion dollars in 2007) are affected by the weather and climate each year; more conservative estimates give the figure of 2.2 trillion dollars annually.[21] Approximately 90% of the presidentially declared disasters in the U.S. are weather-related. As the weather affects countless domains of society, the importance of a better understanding and prediction of the weather is also growing. The space component of this observation and prediction system requires accurate information as well as high-resolution information in both space and time.[22]

Satellites allow the global observation of the atmosphere, the oceans, water, ice and land, which in turn provides more accurate and therefore useful weather and climate information. Instruments onboard these platforms can measure not only ocean levels, ocean salinity, polar ice sheet thickness, oil spills, but also soil humidity, vegetation cover, crop types and evolution, land cover classification, habitat change, etc. Multi-wavelength spectroscopy provides insight into several physical processes and thus yields information on several characteristic parameters of the observed areas. We can carry out interferometer synthetic aperture radar measurements of the topography and derive earthquake-induced land movements at the centimetre scale.[23] The continuity of these observations of the Earth system is however mandatory, as is the need to provide long-term Earth observation data archiving, processing and access facilities to users worldwide. Furthermore, these measurements must be assimilated in models along with other types of measurements. At the European level, the current subdivision of tasks and responsibilities between ESA, national space agencies and the European Commission did not make this target an easy one to achieve but its realisation has been a long-standing request from the scientific community.[24] A first step in the right direction may have just been made with the announcement of the start of the GENESI-DR project funded by the European Commission and led by ESA, taking the lead in providing reliable, easy, and long-term access to Earth science data via the Internet.[25]

Finally, the space component is not only useful in making observations and increasing our knowledge of the Earth system but it also offers the most convenient way to disseminate these observations to users around the globe through an efficient space-based communications system.[26]

As the issue of respect for the environment has become a major concern for the public and, consequently, for governments worldwide, companies are also starting to imbed these concerns into their policies. The concept of Corporate Sustainable Development (CSD) was created a few years ago; there are instances of companies which have already started to incorporate Earth observation data in their CSD portfolio, thus demonstrating their interest in satellite-based data providing a global and cost-effective way to measure objectively the sustainability of business activities.[27]

It is beyond the scope of this paper to present in an exhaustive way the myriad of research areas and applications enabled by space science services that are supporting sustainability on Earth. A more detailed perspective and discussion can be found in ESA 2006, for example.[28] I will only provide four noteworthy examples here.

The first one addresses the major floods experienced in various parts of the world in recent years that have demonstrated the use of space observations for monitoring the evolution of flooded areas and to provide emergency workers in the field with near real-time information, but also to reach a better comprehension of the mechanisms leading from an earthquake to a tsunami. Various satellite-based instruments are being put to use in such instances. In the case of the major tsunami caused by the 9.1–9.3 magnitude undersea earthquake off the west coast of

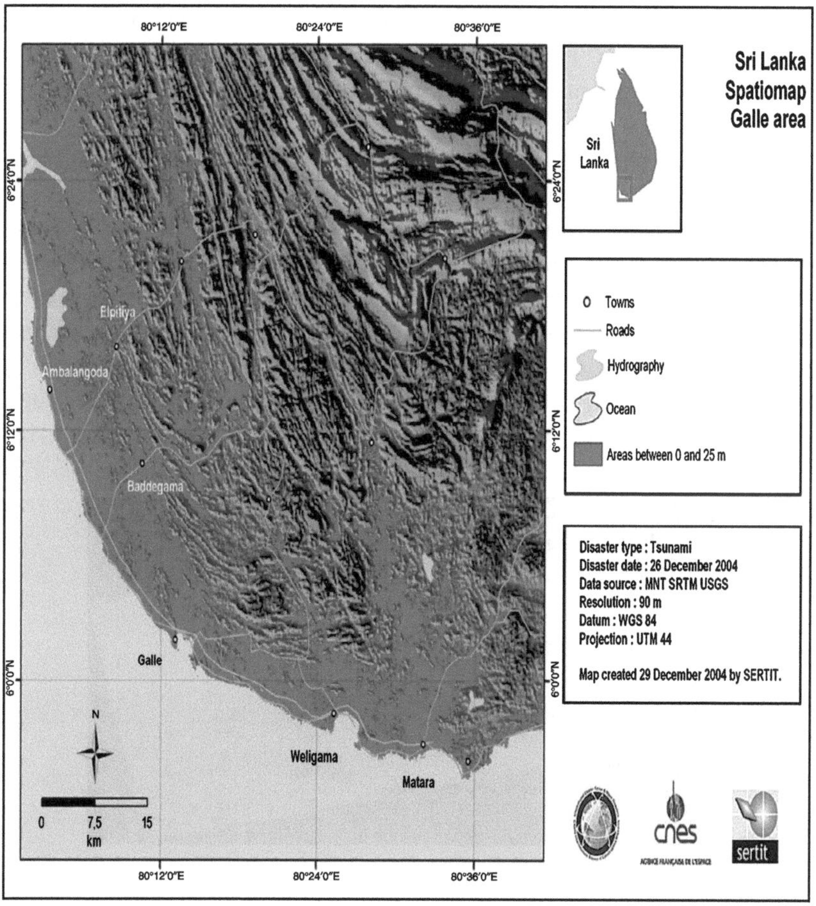

Fig. 6. *Satellite map of lowlands of the Galle area; coast of Sri Lanka in the immediate aftermath of the December 2004 tsunami (source: SERTIT).*

northern Sumatra on 26 December 2004, the satellites LANDSAT-7, SPOT 2, 4 and 5, ERS-2, IKONOS, TERRA (Modis instrument), RADARSAT-1, ENVISAT, and the Space Shuttle Radar Topography Mission (SRTM) provided data about the concerned areas at various wavelengths (visible, cloud-penetrating microwave, imaging spectro-radiometer) and enabled a detailed comparison between data obtained prior to the tsunami and after it. Figure 6 depicts the land portions in Sri Lanka affected by the tsunami.

The Monitoring of land earthquakes has also made enormous progress since the advent of satellite-based radar topography and, in particular, the InSAR (Interferometry Synthetic Aperture Radar) technique that has been used for the past 15 years, allowing researchers to derive earthquake-induced land movements at the centimetre scale.[29] This technique has been put to use as early as 1993 during the Landers earthquake in North America, and has been used widely with various platforms since then. Figure 7 shows such an interferometer map of the 6.6 magnitude Bam earthquake[30] that occurred on 26 December 2003 in the Kerman province in Iran.

A third example concerns the matter of ice caps melting in the polar regions for which the quasi-continuous contribution of Earth observation satellites since the 1970s has allowed researchers to capture the trend. It will also enable them to see whether this trend is reversing, continuing, or worsening. Radar echoes from the ice sheets provide estimates of their elevations. By measuring changes in the elevations of the ice sheets in Antarctica and Greenland, scientists can determine whether they are growing or shrinking and whether global sea levels will be affected.

Researchers are only starting to disentangle the various phenomena occurring at the interface between ice cover, landmass, and sea. Various spectacular events have

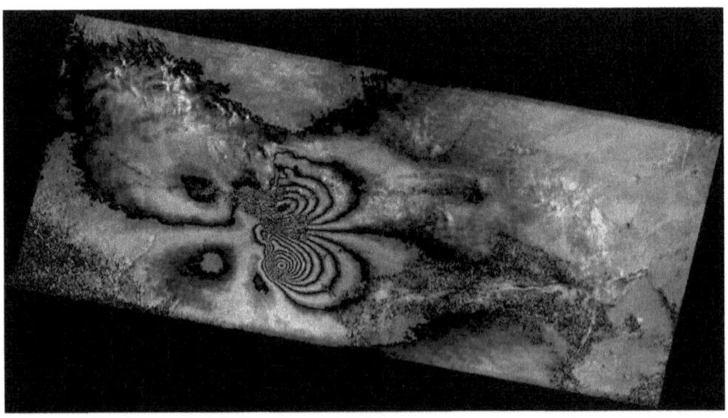

Fig. 7. *This interferogram shows ground motion associated with the 26 December 2003 earthquake at Bam in Iran. The interferogram was created by combining an Envisat Advanced Synthetic Aperture Radar (ASAR) Wide Swath Mode (WSM) image with an Image Mode (IM) image. Processed by Politecnico of Milano (source: Polimi/Poliba).*

been monitored in the past years, such as the recent break-up of the Wilkins ice shelf in Antarctica between 26 February and 7 March 2008, or the fracture of the Larsen-B ice shelf and the drifting of the resulting icebergs into the Weddell Sea in February 2002, both measured by ENVISAT's Advanced Synthetic Aperture Radar (ASAR). ENVISAT carries 10 optical and radar instruments that provide the continuous observation and monitoring of the Earth's land, atmosphere, oceans, and ice caps. The fact is that these events have perhaps taken place since ages and perhaps not. The ignorance of researchers on that point comes from the fact that these satellite-based tools were not available before,[31] which demonstrates the vital use of space-based tools in this domain.

Finally, monitoring the atmosphere is also a domain where space provides substantial added value for observation and data retrieval. One key parameter in terms of global warming is our understanding of the carbon cycle in order to predict its evolution over the next decades. This requires accurate estimates of CO_2 surface fluxes that are currently dependent on data collected by ground-based stations at various locations. The number of these stations is too few and their repartition is not optimal. The answer to this problem once again comes from satellite-based instrumentation that will be able to constantly monitor the atmospheric concentrations of CO_2. Space agencies such as CNES, NASA, and ESA are currently working on the development of measurement concepts able to improve the accuracy of the existing instruments, e.g. the TOVS instrument onboard NOAA's TIROS satellite. The first measurements of CO_2 levels from space were achieved in May 2007 by the German instrument SCIAMACHY onboard ESA's ENVISAT satellite.[32]

Scientific research is fuelling applications and products in an essentially unpredictable way. Despite active research in that area, "... no accurate metrics indeed exist that enable economists to determine both the quality and the future monetary value of economic benefits that may arise from the acquisition of new knowledge".[33] For instance, before the start of the space age, the possible value of satellite data for weather forecasting and other applications was totally unclear, while satellite imagery is now part of our daily life and represents a profitable activity for hundreds of private companies worldwide. Indeed, applications often spring from unexpected areas and unexpected developments. The realisation that basic research must therefore be supported adequately comes with the other realisation that so-called applied research does not exist in isolation.

In the space area, the emergence of whole areas of scientific and technological developments is thus enabled by basic space research. Technological applications in turn fuel scientific developments, and ambitious future operational systems such as GMES and Galileo will therefore also generate further scientific needs and projects.[34]

2.3.5. Other applications

Apart from the basic understanding of the Earth system and the resulting prediction capability for weather, climate, ice coverage, sea level change, and crop yield, space science also supports other fields through which the sustainability of our societies can be addressed, namely in the areas of telecommunications and navigation.

2.3.5.1. Telecommunications

Perhaps the most noteworthy aspect of the use of space assets in everyday life is telecommunications. Communications satellites represent, together with launchers, by far the largest commercial space market worldwide, and the market for satellite communications has been expanding at a sustained rate of more than 15% a year. Over a hundred satellites around the Earth are currently managing radio, television and telephone communications. More than two out of three intercontinental telephone calls are already transmitted by telecommunications satellites, but modern and more powerful satellites as well as the higher frequencies used make it possible for people to receive direct signals from the satellites, for instance with the direct reception of satellite TV programmes.

Information technologies and optronics are the scientific domains enabling the technological development in this area, i.e. in fields such as telemedicine/telesurgery or health alert systems via a satellite network.[35] Many other domains start to benefit from this sector of space, e.g. distributed communication corporate networks or distributed broadcasting and newspaper printing facilities.

2.3.5.2. Navigation

At a not so distant time in the past, travellers looked to the sun and stars to determine their location. Nowadays, our machines operate in the sky regardless of the weather and provide distance references with accuracies ranging from a few centimetres to a few metres. These satellites also provide an extremely accurate time reference by using atomic clocks that keep time within 1 s in 100 million years. Ground-based devices embedding global positioning receivers do not include atomic clocks which are too expensive, so the solution is to let the satellite signals constantly reset the receiver's internal clock. Constellations of such satellites have already been operating for several years. Europe does also have its own fleet ready, with the second launch of the Galileo constellation, Giove-B, in late April 2008.

Navigation satellites thus make heavy use of two main areas not only of modern physics, i.e. relativity and quantum mechanics which provide the tools for developing ever more accurate atomic clocks, but also of electronics and optronics. With the development of affordable cell phones equipped with GPS receivers, many of the innovations being developed in satellite navigation technology will become available to the average citizen. Various projects span a wide range of other applications as well, as in the areas of transport (rail, road, air, maritime), remote asset tracking, emergency management and law enforcement, in-door positioning and navigation, etc.

2.3.6. Watching the skies

As space-based platforms can monitor the Earth in unprecedented detail and continuity, they are of course also studying the rest of the universe, in particular our Sun and the solar system. Indeed, from the standpoint of sustainability on Earth, the only astronomical objects of direct concern are solar flares and meteoroid or asteroid impacts.

2.3.6.1. Space weather

Space weather deals with the changing conditions in the Sun–Earth environment and is believed to be related to the occurrence of solar Coronal Mass Ejections (CMEs). A CME involves the release of a cloud of plasma of approximately 10^{13} kg, carrying an imbedded magnetic field. On arrival on Earth (2 days after the CME occurrence), a magnetic storm can be produced inside the Earth's magnetosphere and this can accelerate particles to very high energies and cause them to be precipitated in regions where spacecraft operate. These particles then can interfere with a spacecraft, usually by changing the bit patterns in the chips that provide memory and on-board computing. These solar storms can also cause magnetic field changes at the surface of the Earth and at high latitudes which can have drastic implications as in the example of the high geo-magnetically induced currents in a Canadian electric power network that caused the destruction of a large transformer supplying the Canadian power grid in 1989. Other effects that are attributed to such solar events are the weakening or breakdown of communication and navigation systems. Another solar storm on 7 January 1997 caused the loss of the 200 million dollar AT&T Telstar 401 telecommunications satellite.

Space weather forecast thus deals with the prediction of outbursts of solar activity, which are of prime importance to satellite operators and ground-based power grid controllers, not to mention their importance for future astronauts en route to Mars. Operational services already exist in the USA[36] and in Europe.[37] Indeed, the origin of CMEs is not yet fully understood and is at present a topic of intense study; a likely explanation is that they could arise due to sudden rearrangements of the magnetic field in the solar corona. The research aims in that domain are thus to become able to recognise unstable magnetic field patterns on the Sun so as to provide a forecast capability of CMEs by studying X-ray images of the corona with spacecraft such as YOHKOH, SOHO, or HINODE.

Space weather-related research and, more broadly, research dealing with the solar influence on the Earth's atmosphere and on climate change have therefore very significant economic and societal implications.

2.3.6.2. Near-Earth Objects (NEOs)

In a series of radio interviews conducted in 2006, Stephen Hawking stated: "Sooner or later disasters such as an asteroid collision or a nuclear war could wipe us all out... But once we spread out into space and establish independent colonies, our future should be safe".[38] Whether the threat of impact by a NEO is as severe as Hawking depicts or not, most commentators agree that "... the threat posed to humanity by NEO impacts is real and similar to other risks of low probability but high consequence which governments take seriously, e.g. earthquakes and volcanic activity".[39] In this case and although the odds are clearly extremely small, the relevance to sustainability on Earth can be as direct as whether we can stay alive as a species.

Is that threat real? How much so? And what can we do in terms of preventive research and mitigation measures? These are obvious questions posed in this case. Scientists and space agencies have started to address the issue, first by trying to establish the threat in quantifiable terms, i.e. by defining classes of objects in terms of size and composition and then by attempting to measure or estimate the population of these objects. Parallel studies of the consequences of impacts at various locations on the Earth and for various classes of objects were carried out. This allowed computing probabilistic estimates of collisions with various kinds of objects and the corresponding consequences of these impacts.

The attention of the media and the general public first turned to the cataclysmic events that could potentially wipe out our civilisation, as is believed to have been the case with the extinction of the dinosaurs and perhaps other mass extinction events throughout the history of the Earth. A recent example of this interest is the observation of the asteroid 99942 Apophis – 2004 MN4[40] which was first presented by the media as having a non-negligible probability of hitting the Earth at its 13 April 2036 pass-by. With currently available orbital data, this probability is of the order of 1 in 8000. Many parameters can be measured and calculated such as possible impact locations, although future passes will most probably change the current orbit. If occurring at sea, the impact of such a 320 m diameter asteroid could create a tsunami potentially as powerful as the one that occurred in December 2004 in South Asia.

Important parameters for estimating the consequences of impacts are the physical properties (composition) of the impacting body. Metal-rich and carbon-rich asteroids will have very different consequences in terms of behaviour in the atmosphere and of energy liberated at the impact site. The physical properties will also affect how the asteroid will respond to tidal forces during successive encounters with the Earth (e.g. 2029 in the case of 2004 MN4), and how it responds to forces, i.e. how we could attach to it and couple energy to divert it.

What kind of science is therefore needed in this context? First of all, a survey of the different classes of objects is mandatory. Remote astronomical observations should be (and are being) performed, measuring spectroscopy, light curves, and polarisation phase curves.[41] This aspect is also relevant to the assessment of space debris. Then, space agencies should envisage space missions to several types of NEOs to measure their structure and composition. Case studies of practical concern should be conducted for advance preparation, assessing such parameters as event probability, warning time, post-warning mitigation, nature of the devastation, but also after-event crisis management for which the contribution of the social sciences is deemed important.[42]

With our current knowledge,[43] mass extinction events seem much too improbable to worry about. Until recently, the greatest threat has been from NEOs with diameters larger than 2 km but that threat is being reduced by the Spaceguard (SG) survey search programme.[44] Much of the remaining threat thus seems to be from "Tunguska"-class impacts, i.e. from objects of a similar size (e.g. a diameter of 30–40 m) as the one which impacted Siberia in 1908. It is estimated that roughly 50% of Tunguska-class NEOs will be found by SG2. The goal is to catalogue 90% of NEOs with diameters larger than 140 m, so it is believed that threatening Tunguska impactors might be detected well in advance.

In practical terms of human psychology and politics, the most likely events are usually of more concern than the unlikely ones, even if they are less destructive. Therefore, public attention has turned to impacts by the smallest albeit most frequent objects.

2.3.6.3. Planetary protection

Finally, one should mention the issue of backward contamination from planetary exploration missions. Life elsewhere in the solar system may not exist; nevertheless, the conduction of scientific investigations of possible extraterrestrial life forms, precursors, and remnants must not be jeopardised and protection measures to this end must be enforced. This is referred to as forward contamination. In addition, and of direct relevance to the subject of this article, Earth must be protected from the potential hazard posed by extraterrestrial matter carried by a spacecraft returning from another planet. Therefore, for certain missions, controls on backward contamination should be imposed and agencies have started to define policies to this end.[45]

2.3.7. Conclusion

Space sciences have proven enormously useful in various domains that are of direct relevance to Earth's sustainability. Although many applications and operational services have already come to fruition, it was argued that research creates applications and products in an essentially unpredictable way and applications generate in turn further scientific needs and projects. To support this virtuous circle, basic science needs to be supported adequately. This is not sufficient, however, and other aspects are mandatory, such as the continuity of data acquisition, its standardisation and proper archiving, and access and distribution for end-users. In Europe, space agencies are playing their role but there is an increased need for direct support from the European Union.

---

[15] European Science Foundation. Humans in Outer Space: Interdisciplinary Odysseys. SCH-ESSC Position Paper. Strasbourg: European Science Foundation, 2008; eds. Codignola, Luca and Kai-Uwe Schrogl. Humans in Outer Space: Interdisciplinary Odysseys. Vienna, New York: Springer, 2009.

[16] Blamont, Jacques. Le Chiffre et le Songe. Paris: Odile Jacob, 1993.

[17] See: Mazur, Peter. "Limits to Life at Low Temperatures and at Reduced Water Contents and Water Activities." Origins of Life and Evolution of Biospheres 10.2 (1980): 137–159; and Mazur, Peter. "Cryobiology: The Freezing of Biological Systems." Science 168 (1970): 939–949; cf. also: Clarke, Arthur C. 2001: A Space Odyssey. New York: New American Library, 1968; and: 2001: A Space Odyssey. Dir. Stanley Kubrick. Metro Goldwyn Mayer, 1968.

[18] Cf. Wilcox, Don. "The Voyage that Lasted 600 Years." Amazing Stories 14.10 (1940); Heinlein, Robert A. "Universe." Astounding Science-Fiction 126 (1941); Adams, Douglas. The Restaurant at the End of the Universe. London: Pan Macmillan, 1980; Russo, Richard P. Ship of Fools. New York: Ace Books, 2001.

[19] Launius, Roger D. and McCurdy, Howard E. Robots in Space: Technology, Evolution and Interplanetary Travel. Baltimore: John Hopkins University Press, 2008.
[20] See: Smith, Lesley J. and Catherine, Doldirina. "Remote Sensing: A Case for Moving Space Data towards the Public Good." Space Policy 24.1 (2008): 22–32; see also: Blamont, Jacques. "We the People: Consequences of the Revolution in the Management of Space Applications." Space Policy 24.1 (2008): 13–21.
[21] Dutton, John A. "Opportunities and Priorities in a New Era for Weather and Climate Services." Bulletin of the American Meteorological Society 83 (2002): 1303–1311.
[22] Committee on NASA-NOAA Transition from Research to Operations; National Research Council. Satellite Observations of the Earth's Environment: Accelerating the Transition of Research to Operations. Washington, D.C.: National Academic Press, 2003.
[23] Massonnet, Didier, et al. "The Displacement Field of the Landers Earthquake Mapped by Radar Interferometry." Nature 364.6433 (1993): 138–142.
[24] See: European Science Foundation. A Strategy for Earth Observation from Space. Strasbourg: European Science Foundation, 1992; see also: European Science Foundation. Recommendations to the ESA Council Meeting at Ministerial Level. Strasbourg: ESF, 2005.
[25] European Space Agency. "ESA Leads Endeavour to Save Earth Science Data." 10 Mar. 2008. http://www.esa.int/esaEO/SEMDFQK26DF_index_0.html
[26] Gall, Robert L., MacDonald, Alexander E., and Serafin, Robert J. "Transition of Weather Research to Operations: Opportunities and Challenges." Bulletin of the American Meteorological Society 83 (2002): 377–391.
[27] European Space Agency. "Satellites Take Sustainability to New Heights." 25 Mar. 2008. http://www.esa.int/esaEO/SEM8P5M5NDF_index_0.html; and: Hatfield Consultants. "Remote Sensing for Sustainable Development Reporting in the Oil Sands." 2008. http://www.hatfieldgroup.com/default.aspx?p=/services/gis_rs/oil+sands.
[28] European Space Agency. The Changing Earth: New Scientific Challenges for ESA's Living Planet Programme. ESA SP-1304. Paris: European Space Agency, 2006.
[29] Massonnet, Didier, et al., 1993 (see endnote 11).
[30] Fialko, Yuri, et al. "Three-Dimensional Deformation Caused by the Bam, Iran, Earthquake and the Origin of Shallow Slip Deficit." Nature 435.7040 (2005): 295–299.
[31] Rapley, Chris. "Understanding the Poles of the Earth, Moon and Mars." Presentation to an SSB-COSPAR Symposium, International Public Seminar Series, Paris, 27 Mar. 2008.
[32] German Aerospace Center. "CO_2 Levels Measured from Space for the First Time by German SCIAMACHY Instrument." 16 May 2007. http://www.dlr.de/en/desktopdefault.aspx/tabid-614/987_read-9092
[33] Hertzfeld, Henry, and Williamson, Ray A. "Socioeconomic Benefits of Earth Science Research." IAF-02-B.5.01. Paper presented at the 53rd International Astronautical Congress, Houston, TX, 18 October 2002.
[34] European Space Agency. "Colloquium on Scientific and Fundamental Aspects of the Galileo Programme." 26 Sept. 2007. http://www.esa.int/esaNA/SEMO9N6H07F_galileo_0.html
[35] European Space Agency. "Medical Care from Space: Telemedicine." 3 Mar. 2008. http://telecom.esa.int/telecom/www/object/index.cfm?fobjectid=8264
[36] See: SpaceWeather.com, http://www.spaceweather.com
[37] See: ESA Space Weather Web Server http://www.esa-spaceweather.net; and: European Space Weather Portal, http://www.spaceweather.eu
[38] Hawking, Stephen. Radio Interviews with BBC (UK) on 30 Nov. 2006 and ABC Radio Australia on 2 Dec. 2006.
[39] European Space Science Committee; European Science Foundation. ESSC-ESF Report on Near-Earth Object Research. Strasbourg: European Science Foundation, 2004; and: UK Task Force. Report on Potentially Hazardous Near-Earth Objects. DTI/Pub 4990/5k/9/00/NP.URN00/1041, 2000.
[40] See: NASA. "99942 Apophis (2004 MN4) Impact Risk." 1 Dec. 2007. http://neo.jpl.nasa.gov/risk/a99942.html

[41] See: Hadamcik, E., et al. "Laboratory Measurements of the Light Scattered by Clouds of Solid Particles by Imaging Technique." Light Scattering Reviews 4: Single Light Scattering and Radiative Transfer. Ed. Alexander A. Kokhanovsky. Berlin: Springer (forthcoming); see also: Worms, Jean-Claude, et al. "Results of the PROGRA2 Experiment: An Experimental Study in Microgravity of Scattered Polarized Light by Dust Particles with Large Size Parameter." Icarus 142 (1999): 281–297.
[42] UNOOSA. Near-Earth Objects Interim Report of the Action Team on Near-Earth Objects (2007–2008). UN Doc. A/AC.105/C.1/L.295, 2008. 14.
[43] Chapman, Clark R. "Meteoroids, Meteors, and the NEO Impact Hazard." Earth, Moon, and Planets 102.1–4 (2008): 417–424.
[44] See: NASA, Near Earth Object Program, http://neo.jpl.nasa.gov/index.html
[45] COSPAR Planetary Protection Policy, approved by the COSPAR Bureau and Council, World Space Congress, Houston, TX, on 20 Oct. 2002. Amended on 24 Mar. 2005.

2.4 Knowledge, data and global sustainability: the role of space-based assets

Mark Williamson

2.4.1. Introduction: knowledge and data

The first satellite, the famous Soviet Sputnik, was launched into space on 4 October 1957, which means the Space Age is now more than half a century old. In those 50 years, mankind has progressed from throwing political balls in the air – to paraphrase Eisenhower's disparaging summary of the Soviet triumph – to operating sophisticated networks of space-based monitoring stations that serve a global community with a wide variety of commercial, civil and military applications. Although satellites have many roles, they could all be said to have one aspect in common: enhancing our knowledge of the Earth and the universe or enabling our access to that knowledge. Knowledge itself is a fuzzy and amorphous concept. In one sense it is simply what we individually or collectively 'know' about a subject, but unfortunately what we think of as knowledge is often only partially understood or incorrectly learned information. This may be a result of the influence of belief systems, preconceived notions or what scientists and engineers call 'bad data,' but it means that our knowledge base is dependent both on viewpoint and the availability of reliable information.

Knowledge, therefore, requires information or data: it is then up to us to decide how to use that data to inform our knowledge on a given topic. When it comes to collecting data about the Earth, its atmosphere, its oceans and its land surfaces, satellites are the perfect tools. This is because satellites offer a global, synoptic view and regular repeat visits as a direct result of the orbital paths they take around the Earth.

Most Earth observation satellites are placed in relatively low altitude orbits (typically between about 600 and 800 km) which take them over the Earth's polar regions. Satellites in these polar orbits have orbital periods of about 90 min, which means that they circumnavigate the planet on a fairly short timescale. Within that orbital period, however, the Earth rotates on its axis by some 22.5°, which means that the satellite takes a different track above the surface on each orbit, returning to the original orbital track after 16 revolutions. Depending on the field of view of the satellite's cameras or other sensors, data from the whole planet can be amassed within a single day, thus providing a potentially large data haul on a repeatable basis.

Specific orbital parameters and narrow fields of view can increase revisit intervals to several days, or even weeks, but the principle of global coverage in a period that would be impossible by terrestrial means (that is, using aircraft) still applies. For

Tab. 1: *Summary of global issues and the role of space-based assets.*

Problems	Solutions
Natural disasters (hurricanes, tsunami, floods, droughts, earthquakes, volcanic eruptions . . .)	Earth observation satellites: data collection (imagery, ground height, etc.); Communications satellites: information dissemination and warnings; Navigation satellites: disaster recovery
Climate instability (global warming, sea level rise, desertification, forest fires)	Earth observation satellites: data
Biodiversity reduction/extinction	Earth observation satellites: data; Navigation satellites: tracking
Population explosion (agricultural stress, water extraction, construction density issues)	Earth observation satellites: crop health, mapping
Environmental degradation/pollution	Earth observation satellites: optical and radar imaging
Uneven wealth distribution (incl. 'digital divide')	Communications satellites: VSATs, DBS, education, health, political information
Global security (military intelligence, surveillance and reconnaissance, anti-terrorism, crime prevention, immigration and border control)	Earth imaging and surveillance/ reconnaissance satellites; Navigation and positioning satellites
Energy scarcity	Earth observation satellites: image analysis for oil and gas prospecting. Potential developments:
	Solar power satellites. Energy relay satellites? Helium 3 lunar mining?
Raw material scarcity	Earth observation satellites: multispectral imaging. Potential developments: Lunar (and asteroid) materials extraction?
NEO impacts (Tunguska events)	Potential developments: Asteroid/comet deflection missions?

example, a type of polar orbit known as a sun-synchronous orbit allows the sub-satellite point to remain approximately fixed at the same local time on Earth, so that the satellite is above a given point at a given time of day (thus maintaining similar lighting conditions for subsequent orbits).

The synoptic and repeatable data collection capabilities of satellites have led to a better understanding of the Earth's weather (and, by extension, climate) simply because it can be imaged, not in isolated parts, but as a complete global system. Indeed, the first photomosaic of the Earth's entire cloud cover was produced as early as February 1965 using 450 pictures from the American Tiros 9 satellite.[46] Thus, weather satellites were arguably the first engine of globalisation.

Following these early developments, it became possible to fly heavier satellites carrying sensors of improved resolution, which brought about developments in the remote sensing of land structures, mineral resources and crop health typified by the Landsat series. Combining the images of weather systems and their effects on land surfaces (such as floods and hurricane damage) allowed satellite data to be used for disaster management and, later, prediction and mitigation.

As shown in the following section and Table 1, Earth observation or Earth imaging satellites have a multitude of uses, as do their military analogues (intelligence, surveillance and reconnaissance satellites). Meanwhile, many different types of science satellites provide images and other data relating to objects and physical processes beyond the confines of the Earth.

Equally important are the satellites that provided the first commercial space application – satellite communications – and those that provide navigation and positioning services. Like Earth observation satellites, both communications and navigation systems operate on a global scale, the former from the high-altitude geostationary orbit (GEO) at 36,000 km, and the latter from medium Earth orbits (MEO) at approximately 19,000 km above the Earth.

Our space-based assets have many roles, providing and distributing large amounts of data on which our knowledge-based society is dependent, but they cannot service all of our needs. Space is certainly not a panacea. We need a combination of space and terrestrial resources to fully meet the requirements of modern-day society. For example, online services such as Google Maps and Google Earth use a combination of satellite and terrestrially-derived data. Meanwhile, in the mobile communications industry, satellite signals (which cannot reach indoors, underground or, unless directly overhead, within the 'canyons' formed by city buildings) are supplemented by a system of ground-based repeaters known as the ancillary terrestrial component (ATC).

Nonetheless, despite their limitations, space-based assets provide a unique perspective and are an indispensable resource (Figure 8).

Fig. 8. *Satellites provide a global view: Earth from Meteosat in geostationary orbit (source: ESA/Eumetsat).*

2.4.2. Satellites and data collection

The sustainability of life on Earth, and by extension the sustainability of our culture and heritage, is threatened by a number of factors that can be addressed by space-based assets. In fact, in a world in which the key issues are 'global issues,' satellites are ideally placed because they provide a global overview. Using the engineering concept of problems and solutions, Table 1 summarises some of these global threats and risks, and the solutions offered by space-based assets.

2.4.2.1. Natural issues

One of the most obvious problems for the world's population – a threat that has existed since human life emerged on this planet – is the natural disaster. This includes atmosphere-related disasters such as hurricanes and tornadoes, ocean-related disasters such as tsunami, and land-based disasters such as floods, droughts, earthquakes and volcanic eruptions.

As mentioned above, Earth observation satellites (including meteorological satellites) have played a key role in providing solutions to these problems since the early days of the Space Age, chiefly by collecting image data. In fact, since

1966, the entire Earth has been photographed at least once a day and, since then, no tropical storm has escaped detection and daily tracking. More recently, following the Asian tsunami of December 2004, the U.S. satellite imaging firms reported the collection of some 80,000 square kilometres of imagery of the area, in the first week alone.[47]

Technology has been developed to detect subtle changes in wave height and overall sea surface levels, which is useful on a short timescale for tsunami monitoring and in the longer term for detecting sea level rises due to thermal expansion and 'global warming.' Similar techniques are used to detect changes in the height of ground surfaces related to tectonic activity (the movement of thrust faults, for example) and the swelling that occurs before some volcanic eruptions. It is hoped that one day, these techniques will allow the prediction of potentially devastating volcanic events, which often have other disastrous consequences such as tsunami.

Moreover, it is important to recognise the role of communications satellites in support of disaster relief and mitigation. Fundamentally, satellites are not affected by terrestrial disasters (unless they rely on a single ground station in the affected region, which is rare). This means that when terrestrial infrastructures are devastated, communications services are still available using satellite phones, distress beacons and other devices. Despite this, analysis in the aftermath of the 2004 Asian tsunami showed that better communications of the impending disaster could have saved many lives, and plans were announced to improve tsunami warning systems.

A category of spacecraft that uses much the same technology as the communications satellite is the navigation satellite, typified by the U.S.-operated Global Positioning System (GPS). For many years, navigation satellites have had a key role in disaster recovery, not least in guiding rescuers in unfamiliar territory where many of the former landmarks may have been obliterated. Positioning satellites have also been used to monitor the movement of land surfaces in the developing field of earthquake prediction.

Another important, and topical, application for Earth observation satellite data is climate change which, since we do not know how permanent this change will be, might better be termed climate instability. This topic evokes the well-known phrases 'global warming' and 'sea level rise,' but also includes local effects such as forest fires and desertification.

Satellites have a key role to play because of their ability to take global snapshots and repeat those snapshots at regular intervals, thereby building a global database of global change. The datasets range from global temperature and radiation balance[48] measurements to ice-cover monitoring and the melting of the polar caps. There is simply no other way to obtain this type of global data on a timescale that is useful.

A third and equally topical use for satellite data arises in the study of biodiversity reduction and extinction of species. At the overview level, imaging satellites can be used to monitor the destruction of natural habitats, which may lead to the extinction of certain species. At the local level, individual animals (of a certain size) can be tracked using navigation satellites to determine how natural or human activities affect migration, for example. This technique was introduced in the early 1970s with NASAs Nimbus satellite series for the unusual application of elk tracking. An electronic collar worn by an animal in the National Elk Refuge of Wyoming was interrogated twice daily by a transceiver payload on Nimbus 3 in a study of migration habits.[49]

2.4.2.2. Human issues

While the degree to which humans are culpable regarding climate change and other forces of nature is open to discussion, there are other threats to sustainability which are decidedly human issues.

One obvious issue is the increase in the world's population. For the poorest nations, the threat tends to be based on the availability of food and water resources, while for nations higher up the development ladder, it is the provision of sustainable employment and welfare services that are paramount.

However, even developed nations have population-related issues where sustainability has not been adequately considered in planning agriculture, housing and transport systems. For example, the challenges in delivering water to large populations in predominantly desert or semi-desert areas has already been dealt with in California by dams, pipelines and irrigation systems. However, it is clear in such areas that water extraction issues and agricultural stress (brought about by over-farming, mineral leaching and inappropriate fertilisation) now threaten the sustainability of the lifestyle these populations have come to expect.

The abilities of the early satellite-borne infrared sensors to monitor the health of crops are well known, but today's multispectral and hyperspectral sensors (which measure a wide range of optical wavelengths) offer greatly enhanced capabilities and much better resolution. Meanwhile, radar satellites can detect small changes in the height of land surfaces that indicate the depletion of aquifers and the lowering of artesian basins as water extraction continues.

The use of satellites for mapping has also evolved considerably since the first satellite images of towns and cities were received to the extent that it is now possible to plan a walking tour of a city centre from a remote computer (which, thanks to satellite communications, can be anywhere in the world). Indeed, combining satellite imagery with computer software allows a multitude of urban planning tasks to be completed in a way that would be impractical or prohibitively expensive without satellites. An example is the mapping of land-use types in urban areas, which shows an obvious increase in hard, non-porous surfaces, such as roads and car parks, at the expense of natural, porous surfaces, such as grass and scrubland. This increases rainfall run-off, as opposed to absorption, which without adequate drainage can lead to flooding.

A key effect of increasing population density, coupled of course with modern-day lifestyle issues, is an increase in environmental degradation, and more specifically pollution. Apart from the abilities of imaging satellites to monitor the reduction of land area available to agriculture and the growth of landfill sites, different satellite sensors are able to monitor the development and chemical constituents of airborne pollution on a global scale and detect pollution in river estuaries and oceans. Indeed, one of the early demonstrations of radar satellite capabilities was the detection of illegal tank-flushing by oil tankers at sea, made possible because water covered by oil appears flat on the images.

Communications satellites, too, have a role to play in the mitigation of many serious human issues, such as poor health, education and the uneven distribution of wealth. The first serious experiment in this regard was conducted by India using NASA's Applications Technology Satellite, ATS-6, in the mid-1970s. It broadcast educational TV to the rural populations of India via 3-m chicken-wire antennas placed in thousands of villages. As a result of this application, which later became an industry, ATS-6 could be considered the first direct

Fig. 9. *VSATs are everywhere (a hotel roof in Hyderabad) (source: Mark Williamson).*

broadcasting satellite and the first to conduct distance education on an international scale.

Now that it is possible to install so-called VSATs (very small aperture terminals) almost anywhere in the world (Figure 9), education and health information, and of course political information, can be transmitted directly to even the smallest, most remote communities. And with the rise of the Internet, portable computers and mobile satcoms systems, the communications satellite is well placed to bridge the 'digital divide.'

Beyond the civilian and commercial sphere, there are of course other human issues which require the intervention of military and security forces, and their dedicated imaging and communications systems. At one level are the classical military tasks (intelligence, surveillance and reconnaissance) and, at another, are global security issues such as anti-terrorism, crime prevention, immigration and border control. Imaging, navigation and communications satellites have an ever increasing role to play in collecting, locating and communicating data for these applications.

2.4.2.3. Future issues

Thus far, we have considered issues that are largely historical in that they have been known and understood for many years and have already benefited from space-based solutions. Moreover, the relevant space-based assets are all in orbit around the Earth. However, the potential exists to extend the operational realm beyond the Earth's orbit to support Earth-based problems.

The first issue, continually highlighted as a key concern for the future, is energy scarcity. Indeed, many potential Earth-based solutions are already under development, from wind and wave farms to new-build nuclear plants. Moreover, Earth Observation satellites have been used for many years for oil and gas prospecting, and geologists are now experienced in analysing images that indicate the presence of new resources.

However, some researchers believe that this is only a stop-gap and that the ultimate energy supply solution involves tapping energy directly from the Sun. Most terrestrially minded people think immediately of paving the deserts with solar panels, but since Peter Glaser popularised the concept of the solar power satellite in the late 1960s,[50] space professionals have been discussing its feasibility and looking forward to the day when it becomes cost-effective.

In a nutshell, this would involve building very large arrays of solar cells in the Earth's orbit that convert the electrical energy they produce into microwave energy and beam it down to large collectors on the Earth's surface, whereupon it would be reconverted into electrical energy and fed into grids. An associated space-based system that has been proposed is the energy relay satellite: a network of such satellites would transmit microwave energy around the globe and beam it down to its point of use (presumably more efficiently than distribution via terrestrial high-voltage power lines). Although the case for such systems remains unproven, the basic technology is available and space-based assets could have a role to play in the field of energy generation in future.

Somewhat more 'pie in the sky' are proposals to mine the lunar surface for helium-3 deposited by the Sun in the upper layers of the lunar soil or 'regolith.' The helium-3 would then be transported back to Earth to power as-yet-unbuilt nuclear fusion reactors to produce more-or-less 'unlimited power.' At best, this space-based solution to Earth's energy problems is a long-term one; at worst, it is simply a scheme dreamed up by space proponents to stimulate the development of the infrastructure required to colonise and industrialise the Moon. Although it is beyond the scope of this paper, some specialists argue that helium-3 would not provide the 'clean energy' solution proposed, thereby eliminating the advantage of lunar helium mining, so this is very much a work in progress.

A second future global issue is the growing scarcity of important raw materials on Earth, not least those required for the all-pervasive electronics industry. Again, multispectral imaging from Earth observation satellites has been employed for many years to identify mineral content and potential sites for mining. Given the required technology, the obvious extension to this is to survey and potentially extract materials from the Moon or lesser celestial bodies such as comets and asteroids. The technology required to image and land on these bodies has already been proven, but the business case for surveying and mining any of these bodies remains absent. One has to assume that the supply situation on Earth will have to get much worse before funding is made available to develop such missions and the related infrastructure that would be needed.

A final example of the potential role of space-based assets in ensuring the sustainability of life on Earth is, ostensibly, even closer to science fiction than solar power satellites and lunar mining. It is, however, evoking serious interest among scientists and governments alike, because it could be the ultimate threat to sustainability.

Near Earth Objects, or NEOs, are comets, asteroids or large meteoroids whose orbits around the Sun bring them close to the Earth, potentially close enough to enter the atmosphere and impact the surface. There are several sites on Earth thought to be the result of such impacts and they are even implicated in the extinction of the dinosaurs. The famous Tunguska Object, which decimated a Siberian forest on 30 June 1908, is believed to have been an asteroid which exploded on entering the atmosphere.

In recent years, space agencies and industrial companies have undertaken design studies for spacecraft that could land on such objects before they come too close to the Earth and deflect them from their destructive courses. Even the UK government – not known for its flights of fancy regarding space exploration – commissioned a study of the NEO threat in 2000.

The multiple impacts of Comet Shoemaker-Levy 9 on Jupiter, observed in 1995, gave cause for thought, while supporting evidence for the threat came in 2002 with the detection of a 'football-field-sized object' that passed between the Earth and the Moon, just 120,000 km away – a near miss on the planetary scale. It could be just a matter of time before the technology for NEO deflection becomes a necessity.

But if all of this sounds too much like science fiction to be true, one should consider the predictive power of that genre. When the original *Star Trek* TV series was released in the mid-1960s, the hand-held 'communicators' used by the crew to contact their spacecraft (in a stationary orbit about the planet) were far beyond the technology of the day. But the world did not have to wait until the 23rd century for communications technology to catch up: the hand-held satellite phone became reality before the end of the 20th century.

When it comes to the application of space-based assets, it is wise to remain open-minded.

2.4.3. Ensuring continuity

Although space technology has proved itself useful in mitigating threats to the environment, and sustainability issues in general, the sustainability of the technology itself is also under threat.

The choice of the term 'space-based assets' is deliberate, because the sustainability of our satellite infrastructure is dependent on the recognition of 'asset value' in a business and financial sense. This has been achieved for our space-based communications infrastructure, because satellites are ordered, financed, operated and insured on a commercial basis. If they fail in orbit, there are financial as well as operational repercussions. Satellites appear as assets on the operators' balance sheets and have a definable asset value.

Although this is also the case for the commercially operated imaging satellites, it is not so for most government Earth imaging, meteorology and science satellites. They are not insured for their in-orbit lives and are therefore not renewable assets should they fail.

This has led to a data continuity problem, which has become even more critical in the light of the current interest in long-term climate data. If a given satellite, or payload sensor on that satellite, fails, there is rarely a financial mechanism available to replace that resource, because political funding is not long-term. A case in point is the Landsat Earth observation programme which, despite its success, has been starved of funds to the point where the remaining Landsat spacecraft are either past their design lives (and therefore prone to immediate failure) or have suffered sensor anomalies which degrade their data.

A related problem is that payloads are highly specialised and other sensors in orbit are incapable of substituting for them. Data users in the scientific community have become used to receiving this specific data and, since their theories and careers are often based on such data, a failure can affect much more than a satellite. Despite this, there is no long-term plan for data continuity.

Meteorology was the first Earth observation application to be recognised as having value – its effects have obvious implications for the insurance industry, for example – but for Europe's Meteosat system, by some estimates, the transition from scientific curiosity to commercial asset took at least 15 years. The development with regard to other applications is still open to question: for example, what is the value of monitoring sand blown from the Sahara across the Atlantic, where it fertilises the soil of the Amazon basin? The value of monitoring sea surface heights, glacier calving and the ozone hole should be obvious, but the lack of guaranteed funding to sustain the space-based assets capable of doing so suggests otherwise.

There are many reasons why sufficient funding is not readily available for space-based systems, one of which relates to the relatively small size of the space community and its relatively small voice compared with terrestrial providers and proponents. But it is also due to the space community's insufficient commitment to public relations, marketing and educational outreach. Take, for example, the lack of visibility of the European Space Agency in one of its member states, the UK, where the majority of the population would not recognise the acronym ESA. By contrast, in the same country, NASA is a byword. This is partly due to the Apollo lunar missions, which galvanised a generation, but also because of publicity surrounding the Space Shuttle, Hubble Space Telescope and other high-profile missions. Indeed, NASA is not just a familiar acronym; it is even accorded the familiarity of a proper name – written "Nasa" – by some of the general media. ESA and the European space industry should take note.

Finally, in our analysis of sustainability, we should consider the continuity of the resource on which space-based assets rely, that of space itself: specifically, the resources of orbital space from which to operate the satellites, and frequency space by which we communicate and transfer data.

The limited amount of frequencies for the transmission and reception of data and other communications products is coordinated at the international level by the

International Telecommunication Union (ITU) and certain parts of the frequency spectrum are recognised as commercial assets (and sometimes auctioned off). The use of geostationary orbital positions and other orbits is also coordinated by the ITU, but the recognition of their asset value and the consequent need to protect them for future sustainable use has taken longer to realise.[51]

For perhaps half of the 50 years of the Space Age, those who foresaw the problems of uncontrolled orbital debris – the result of exploding satellites and launch vehicle stages, for example – failed to convince agencies and operators that the growth in debris threatened the sustainability of the orbital resource (Figure 10). In recent years, however, the problem has become too great to ignore and the need for debris mitigation measures is widely recognised (specifically by guidelines proposed by the Inter-Agency Space Debris Coordination Committee (IADC) and adopted in February 2007 by the UN General Assembly). Nonetheless, the deliberate destruction of a Chinese weather satellite in a Chinese anti-satellite weapon test of January 2007, which produced thousands of new debris objects, illustrates the continuing reality of the threat.

Space is recognised in U.S. and European space policy as part of the Earth's critical infrastructure, while certain coordinating bodies recognise orbital and frequency space as assets worthy of protection. For these reasons alone, these

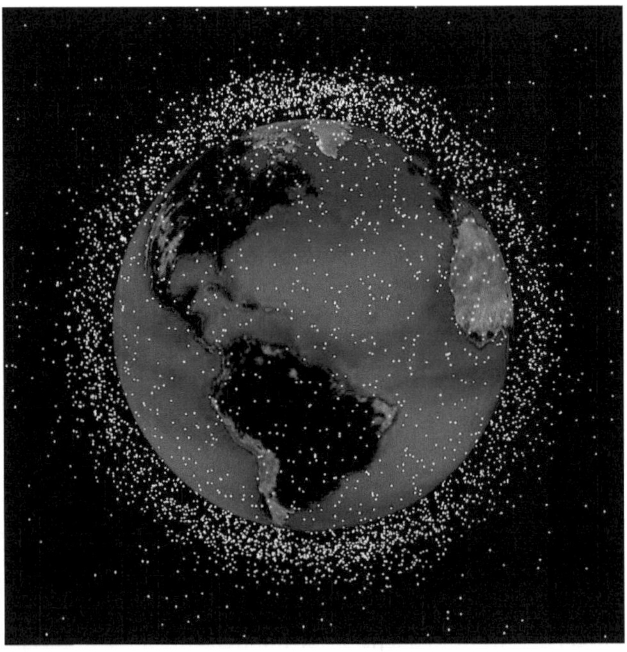

Fig. 10. *Computer-generated graphic of tracked objects in low Earth orbit (LEO) (source: NASA).*

natural space assets should be recognised as an integral part of the 'Earth system' and thus part of the global sustainability issue.

The 1987 Brundtland Report[52] defined sustainable development as "development that meets the needs of the present without compromising the ability of future generations to meet their own needs". Although this was applied to the terrestrial environment, it is clear that it is readily transferable to the space environment. Unfortunately, the sustainability of the space environment is not sufficiently protected for future generations by existing space law treaties, which are state-based. The actions of individuals, such as space development entrepreneurs and space tourists, are not covered.

Space-based assets have a role as guardians of sustainability on Earth. We must hope that agreement on protecting the natural assets of the space environment is reached before they become unusable.

2.4.4. Conclusions

The use of space-based assets, such as Earth observation and communications satellites, shows clearly that space applications help to address global sustainability issues by providing a unique perspective of the planet and, as such, offer an indispensable resource for the prediction, management and mitigation of natural and man-made disasters and other issues. It is also clear, however, that more must be done to maintain and enhance these in-orbit resources to ensure the continuity of important data, that is, by ensuring the continued funding of programmes and intellectual resources. As a step towards reducing duplication and potential wastage of resources, we should also encourage the establishment and funding of international networks and data sharing. Finally, since space-based assets have a crucial role to play in monitoring threats to global sustainability, we should also take action to ensure the sustainability of the space environment itself.

---

[46] Williamson, Mark. Spacecraft Technology: The Early Years. London: Institution of Electrical Engineers, 2006. p. 136.
[47] Williamson, Mark. "Catch the Wave." Institution of Electrical Engineers Review 51 (2005): 30–34.
[48] The balance between incoming solar energy and outgoing thermal radiation.
[49] Turnill, Reginald, ed. Jane's Spaceflight Directory. London: Jane's Publishing Company, 1987. p. 84.
[50] Glaser, Peter E. "Power from the Sun: Its Future." Science Magazine 162 (1968): 857–861.
[51] Williamson, Mark. Space: The Fragile Frontier. Reston: American Institute of Aeronautics and Astronautics, 2006. p. 43.
[52] United Nations. Our Common Future. Report of the World Commission on Environment and Development. Published as Annex to General Assembly Doc. A/42/427 "Development and International Co-operation: Environment." 2 Aug. 1987.

CHAPTER 3

ENVIRONMENT

3.1 Overview

The context: It has become evident that drastic changes have occurred in the environment in the last few decades. Humanity has been faced with natural disasters which – together with global warming – constantly deteriorate the environment humans have become accustomed to living in. All this can lead not only to changes in lifestyles around the world but should also prompt humanity to learn and develop means which would mitigate and offer possible solutions to the issues at stake.

The contribution of space: Space provides very accurate data measurements, recorded and transmitted independently by numerous satellites. Such data are a lot more precise in comparison to data measurements done on Earth. From space, the global monitoring of carbon dioxide, methane and other gases can easily be done. Also, the melting of the polar ice cups can be accurately assessed through space technology. Moreover, natural disasters can be monitored from space and managed through initiatives such as the International Charter "Space and Major Disasters" or programmes such as Global Monitoring for Environment and Security (GMES).

The papers: The papers in this section offer a general overview of the current environmental situation on Earth and discuss the possible contributions which space could make to the improvement of that situation.

In "Climate Change and Sustainability", Charlotte Mathieu reviews the challenges associated with climate change that humankind is facing and should address. The paper is based on the presentation of the IPCC Fourth Assessment Report by IPCC Chairman R.K. Pachauri in Valencia, Spain on 17 November 2007, as well as on the Fourth Assessment Report "Climate Change 2007: Synthesis Report" and its "Summary for Policymakers". Besides discussing the causes for climate change and its effects on the environment, the paper also outlines some projections for the upcoming decades and possible mitigation options.

The paper "The Contribution of Earth Observation to Environmental Monitoring and Global Monitoring for Environment and Security (GMES)" was written by Pierre Potin and Josef Aschbacher from the GMES Space Office at the Directorate of Earth Observation Programmes of the European Space Agency. It is divided into two sections: the first section describes the contributions of Earth observation to environmental monitoring, presenting scientific results and applications which use data from ERS and ENVISAT missions. The second section gives a detailed account of the GMES initiative and the programmatic aspects of its Space Component.

In the last paper entitled “Environmental Threats and Space Responses”, Jérôme Lafeuille from the World Meteorological Organization’s Space Programme discusses the many environmental threats which are likely to worsen in the coming decades. He points out the unique role which space technology, and particularly Earth observation, could play in the management and mitigation of those threats. For instance, he argues that there are many reasons why the space-based Global Observation System should be expanded towards a comprehensive monitoring system for the climate and the environment.

3.2 Climate change and sustainability

Charlotte Mathieu

The Intergovernmental Panel on Climate Change (IPCC) is an intergovernmental, scientific body established by the WMO and the UN Environment Programme in 1988. It provides decision-makers with an independent and objective source of information on climate change, its causes, its potential consequences, and the adaptation and mitigation options for responding to it.

This paper reviews the challenges associated with climate change that humankind is facing and should address. It is based on the presentation of the IPCC Fourth Assessment Report by IPCC Chairman Dr. R.K. Pachauri in Valencia, Spain on 17 November 2007, as well as on the Fourth Assessment Report "Climate Change 2007: Synthesis Report" and its "Summary for Policymakers".

3.2.1. Observed changes in the climate and their effects

The warming of the climate system is unequivocal, as demonstrated by increasing global and air temperatures, a rising global average sea level, and the widespread melting of snow and ice (see Figure 1).

The frequency of, and the area affected by, extreme events have increased over the past decades. It is likely that the frequency of heat waves has increased over most land areas and that heavy precipitations have become more frequent in most areas. From 1900 to 2005, precipitation increased significantly in the eastern parts of North and South America, northern Europe and central Asia. However, it declined in the Sahel zone, the Mediterranean, southern Africa and parts of southern Asia. Globally, the area affected by droughts is likely to have increased since the 1970s.

Consistent with warming is the increasing sea level rise. The rate of the global average sea level rise increased by almost 75% between 1961 and 1993. This resulted from the combined effects of thermal expansion and the melting of glaciers, ice caps, and the polar ice sheets. The projected sea level rise for the whole 21st century ranges from 18 to 59 cm.

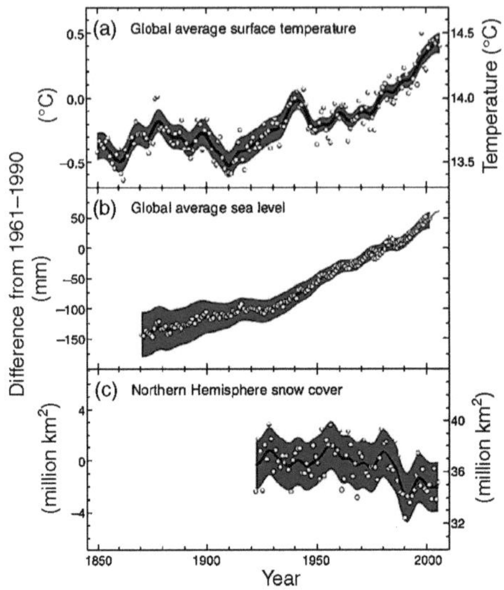

Fig. 1. *Changes in temperature, sea level, and Northern Hemisphere snow cover.*[53]

3.2.2. Causes of those changes

Humans have contributed to this climate change. Their activities have influenced the climate, and specifically temperatures. There is very high confidence that the

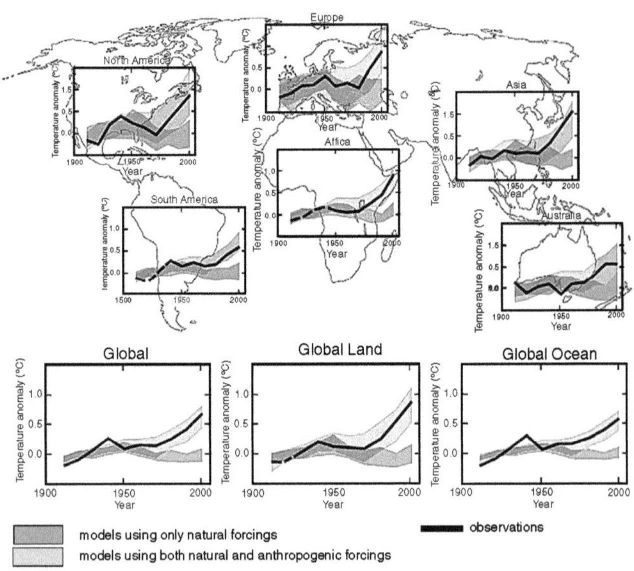

Fig. 2. *Estimated anthropogenic contributions to global and continental temperature change.*[53]

net effect of human activities since the mid-18th century has been one of warming. A large share of the observed increase in global temperatures since the 1950s is very likely due to the increase in the atmospheric concentrations of anthropogenic greenhouse gases. As illustrated by Figure 2, it is likely that there has been significant anthropogenic warming over the past 50 years.

In addition, human activities have likely impacted other characteristics of the climate. Humans have probably contributed to sea level rise, the change of wind patterns, and increased temperatures (extremely hot and cold nights, and cold days).

3.2.3. Projections

Much evidence supports the hypothesis that with the current situation, global greenhouse gas emissions will continue to grow over the next few decades, which will cause further warming and induce many changes in the global climate system in this century. Figure 3 shows the projected surface warming for this century.

Greenhouse gas emissions at or above the current rates will cause changes that are very likely to be larger than those observed during the 20th century. Even if the concentrations of those gases were to be stabilised, anthropogenic warming and sea level rise will continue for centuries because of the time scales associated with climate processes and feedbacks.

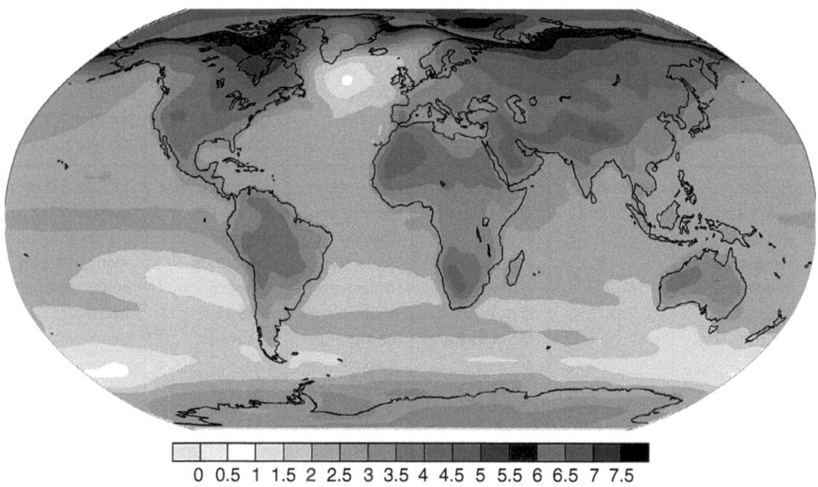

Fig. 3. *Geographical pattern of surface warming – projected changes between the late 20th century and the late 21st century.*[53]

Tab. 1: *Examples of possible impacts of climate change due to changes in extreme weather and climate events.*[53]

Phenomenon[a] and direction of trend	Likelihood of future trends based on projections far 21st century using SRES scenarios	Examples of major projected impacts by sector			
		Agriculture, forestry and ecosystems	Water resources	Human health	Industry, settlement and society
Over most land areas, warmer and fewer cold days and nights, warmer and more frequent hot days and nights	*Virtually certain*[b]	Increased yields in colder environments; decreased yields in warmer environments; increased insect outbreaks	Effects on water resources relying on snowmelt; effects on some water supplies	Reduced human mortality from decreased cold exposure	Reduced energy demand for heating; increased demand for cooling; declining air quality in cities; reduced disruption to transport due to snow, ice; effects on winter tourism
Warm spells/ heatwaves. Frequency increases over most land areas	*Very likely*	Reduced yields in warmer regions due to heat stress; increased danger of wildfire	Increased water demand; water quality problems, e.g. algal blooms	Increased risk of heat-related mortality, especially for the elderly, chronically sick, very young and socially isolated	Reduction in quality of life for people in warm areas without appropriate housing; impacts on the elderly, very young and poor
Heavy precipitation events. Frequency increases over most areas	*Very likely*	Damage to crops: soil erosion, inability to cultivate land due to waterlogging of soils	Adverse effects on quality of surface and ground-water; contamination of water supply; water scarcity may be relieved	Increased risk of deaths, injuries and infectious, respiratory and skin diseases	Disruption of settlements, commerce, transport and societies due to flooding; pressures on urban and rural infrastructures; loss of property

(continued)

Tab. 1: (*continued*)

Phenomenon[a] and direction of trend	Likelihood of future trends based on projections far 21st century using SRES scenarios	Examples of major projected impacts by sector			
		Agriculture, forestry and ecosystems	Water resources	Human health	Industry, settlement and society
Area affected by drought increases	*Likely*	Land degradation; lower yields/ crop damage and failure; increased livestock deaths; increased risk of wildfire	More widespread water stress	Increased risk of food and water shortage; increased risk of malnutrition; increased risk of water- and food-borne diseases	Water shortage for settlements, industry and societies; reduced hydropower generation potentials; potential for population migration
Intense tropical cyclone activity Increases	*Likely*	Damage to crops; wind-throw (uprooting) of trees; damage to coral reefs	Power outages causing disruption of public water supply	Increased risk of deaths, injuries, water- and food-borne diseases; post-traumatic stress disorders	Disruption by flood and high winds; withdrawal of risk coverage in vulnerable areas by private insurers; potential for population migrations; loss of property
Increased incidence of extreme high sea level (excludes tsunamis)[c]	*Likely*[d]	Salinisation of irrigation water, estuaries and fresh-water systems	Decreased freshwater availability due to saltwater intrusion	Increased risk of deaths and injuries by drowning in floods; migration-related health effects	Costs of coastal protection versus costs of land-use relocation; potential for movement of populations and infrastructure; also see tropical cyclones above

The altered frequencies and intensities of extreme weather, together with sea level rise, are expected to have mostly adverse effects on natural and human systems, as illustrated in Table 1.

Human influences on climate warming could have some irreversible impacts. The partial loss of ice sheets on polar ice land could imply metres of sea level rise, major changes in coastlines, or the inundation of low-lying areas (with the greatest effects in river deltas and low-lying islands). There is medium confidence that 20–30% of the species assessed so far are likely to be at an increased risk of extinction. The Meridional Overturning Circulation (MOC) of the Atlantic Ocean will very likely slow down during this century. The possible large-scale and persistent changes in the MOC are likely to have an impact on marine ecosystem productivity, fishery, ocean carbon dioxide uptake, oceanic oxygen concentrations, and terrestrial vegetation. Changes in carbon dioxide uptake may then affect the climate.

3.2.4. Adaptation and mitigation options

There are many options available to adapt to the ongoing climate change, but more extensive adaptation is required in order to reduce humanity's vulnerability to climate change. In terms of mitigation, much evidence indicates that there is substantial potential for the mitigation of global carbon dioxide emissions over the coming decades that could offset the projected growth of global emissions or even reduce emissions below the current levels.

Table 2 details the key mitigation technologies and practices currently available (or to be available before 2030) in various sectors, and the associated policies which are deemed to be environmentally effective.

A wide array of polices and instruments are available to governments for creating the incentives for mitigation action. They include the integration of climate policies into wider development policies; regulations and standards, taxes and charges, tradable permits, financial incentives, voluntary agreements, information instruments, and research and development.

Models demonstrate that certain levels of the carbon price, expressed as costs per unit of carbon dioxide-equivalent emission avoided or reduced, have the potential to significantly impact emissions in all sectors. An effective carbon-price signal has the potential to realise significant greenhouse gas reductions.

Making development more sustainable can enhance mitigation capacities and reduce emissions. However, it is very likely that climate change can slow the pace of progress towards sustainable development.

Tab. 2: *Examples of key mitigation technologies and policies in selected sectors.*[53]

Sector	Key mitigation technologies and practices currently commercially available. *Key mitigation technologies and practices projected to be commercialised before 2030 shown in italics*	Policies, measures and instruments shown to be environmentally effective	Key constraints or opportunities (Normal font = constraints; *italics = opportunities*)
Energy supply	Improved supply and distribution efficiency; fuel switching from coal to gas; nuclear power; renewable heat and power (hydropower, solar, wind, geothermal and bioenergy); combined heat and power; early applications of carbon dioxide capture and storage (CCS) (e.g. storage of removed CO_2 from natural gas); CCS for *gas, biomass and coal-fired electricity generating facilities; advanced nuclear power; advanced renewable energy, including tidal and wave energy, concentrating solar*, and *solar photovoltaics*	Reduction of fossil fuel subsidies; taxes or carbon charges on fossil fuels Feed-in ariffs for renewable energy technologies; renewable energy obligations; producer subsidies	Resistance by vested interests may make them difficult to implement *May be appropriate to create markets for low-emissions technologies*
Transport	More fuel-efficient vehicles; hybrid vehicles; cleaner diesel vehicles; biofuels; modal shifts from road transport to rail and public transport systems; non-motorised transport (cycling, walking); land-use and transport planning; *second generation biofuels; higher efficiency aircraft; advanced electric and hybrid vehicles with more powerful and reliable batteries*	Mandatory fuel economy; biofuel blending and CO_2, standards for road transport Taxes on vehicle purchase, registration, use and motor fuels; road and parking pricing Influence mobility needs through land-use regulations and infrastructure planning; investment in attractive public transport facilities and non-motorised forms of transport	Partial coverage of vehicle fleet may limit effectiveness Effectiveness may drop with higher incomes *Particularly appropriate for countries that are building up their transportation systems*

(continued)

Tab. 2: *(continued)*

Sector	Key mitigation technologies and practices currently commercially available. *Key mitigation technologies and practices projected to be commercialised before 2030 shown in italics*	Policies, measures and instruments shown to be environmentally effective	Key constraints or opportunities (Normal font = constraints; *italics = opportunities*)
Buildings	Efficient lighting and day lighting; more efficient electrical appliances and heating and cooling devices; improved cook stoves, improved insulation; passive and active solar design for heating and cooling; alternative refrigeration fluids, recovery and recycling of fluorinated gases; *integrated design of commercial buildings including technologies, such as intelligent* meters *that provide feedback and control; solar photovoltaics integrated in buildings*	Appliance standards and labelling Building codes and certification	Periodic revision of standards needed *Attractive for new buildings* *Enforcement can be difficult*
		Demand-side management programmes	Need for regulations so that utilities may profit
		Public sector leadership programmes, including procurement	*Government purchasing can expand demand for energy-efficient products*
		Incentives for energy service companies (ESCOs)	*Success factor: Access to third party financing*
Industry	More efficient end-use electrical equipment; heat and power recovery; material recycling and substitution; control of non-CO_2 gas emissions; and a wide array of process-specific technologies; *advanced energy* efficiency; CCS for *cement,* ammonia, *and iron manufacture; inert electrodes for aluminium manufacture*	Provision of benchmark information; performance standards; subsidies; tax credits	*May be appropriate to stimulate technology uptake.* Stability of national policy important in view of international competitiveness
		Tradable permits	Predictable allocation mechanisms and stable price signals important for investments
		Voluntary agreements	Success factors include: clear targets, a baseline scenario, third-party involvement in design and review and formal provisions of monitoring, close cooperation between government and industry

(continued)

Tab. 2: (*continued*)

Sector	Key mitigation technologies and practices currently commercially available. *Key mitigation technologies and practices projected to be commercialised before 2030 shown in italics*	Policies, measures and instruments shown to be environmentally effective	Key constraints or opportunities (Normal font = constraints; *italics = opportunities*)
Agriculture	Improved crop and grazing land management to increase soil carbon storage; restoration of cultivated peaty soils and degraded lands; improved rice cultivation techniques and livestock and manure management to reduce CH_4 emssions; improved nitrogen fertiliser application techniques to reduce N_2O emissions; dedicated energy crops to replace fossil fuel uses improved energy efficiency; *improvements of crop yields*	Financial incentives and regulations for improved land management maintaining soil carbon content; efficient use of fertilisers and irrigation	*May encourage synergy with sustainable development and with reducing vulnerability to climate change, thereby overcoming barriers to implementation*
Forestry/ forests	Afforestation; reforestation; forest management; reduced deforestation; harvested wood product management; use of forestry products for bioenergy to replace fossil fuel use; *tree species improvement to increase biomass productivity and carbon sequestration improved remote sensing technologies for analysis of* vegetation/soil *carbon sequestration potential and mapping land-use change*	Financial incentives (national and international) to increase forest area, to reduce deforestation and to maintain and manage forests; land-use regulation and enforcement	Constraints include lack of investment capital and land tenure issues. *Can help poverty alleviation*
Waste	Landfill CH_4 recovery; waste incineration with energy recovery; composting of organic waste; controlled wastewater treatment recycling and waste minimisation; biocovers *and biofilters to optimise CH_4 oxidation*	Financial incentives for improved waste and wastewater management Renewable energy incentives or obligations Waste management regulations	*May stimulate technology diffusion* Local availability of low-cost fuel Meet effectively applied at national level with enforcement strategies

3.2.5. The long-term perspective

According to Article 2 of the UN Framework Convention on Climate Change (UNFCC), the objective of the UNFCC is to achieve the "stabilization of greenhouse gas concentrations in the atmosphere at a level that would prevent dangerous anthropogenic interference with the climate system". Determining what constitutes "dangerous anthropogenic interference" involves value judgments, but science can support informed decisions on this issue.

Neither adaptation nor mitigation endeavours alone can avoid all climate change impacts, but together they can significantly reduce the risks of climate change. Many impacts can be reduced, delayed or even avoided by mitigation. Mitigation efforts over the next couple of decades will have a significant impact on our opportunities to achieve lower stabilisation levels. Later emission reductions will largely constrain the opportunities to achieve lower stabilisation levels and increase the risk of more severe climate change impacts. This is illustrated by the example of sea level rise which is inevitable under warming, at least because of thermal expansion. Thermal expansion would continue for centuries after greenhouse concentrations have stabilised, and as illustrated in Table 3, the sooner the

Tab. 3: *Long-term global average sea level rise due to thermal expansion, only for the scenarios assessed.*[53]

Category	CO_2 concentration at stabilisation (2005 = 379 ppm)[b]	CO_2 equivalent concentration at stabilisation including GHGs and aerosols (2005=375 ppm)[b]	Peaking year for CO_2 emissions[a, c]	Change in global CO_2 emissions in 2050 (percent of 2000 emissions)[a, c]	Global average temperature increase above pre-industrial at equilibrium, using 'best estimate' climate sensitivity[d, e]	Global average sea level rise above pre-industrial at equilibrium from thermal expansion only[f]	Number of assessed scenarios
	ppm	ppm	Year	Percent	°C	Metres	
I	350–400	445–490	2000–2015	−85 to −50	2.0–2.4	0.4–1.4	6
II	400–440	490–535	2000–2020	−60 to −30	2.4–2.8	0.5–1.7	18
III	440–485	535–590	2010–2030	−30 to +5	2.8–3.2	0.6–1.9	21
IV	485–570	590–710	2020–2060	+10 to +60	3.2–4.0	0.6–2.4	118
V	570–660	710–855	2050–2080	+25 to +85	4.0–4.9	0.8–2.9	9
VI	660–790	855–1130	2060–2090	+90 to +140	4.9–6.1	1.0–3.7	5

emission are stabilised, the lower the stabilisation levels, and the more limited the consequences will be (in the example of global average sea level rise only due to thermal expansion).

Much evidence supports the hypothesis that all stabilisation levels assessed can be achieved through the deployment of a portfolio of technologies that are either currently available or expected to be commercialised in the coming decades.

3.2.6. Equity issues

As a final point, it is important to underline that the impacts of climate change will not be distributed uniformly over the globe. In Africa by 2020, between 75 and 250 million people are projected to be exposed to increased water stress, and in some countries, yields from rain-fed agriculture will be reduced by 2020. In Asia, freshwater availability is projected to decrease by the 2050s, and coastal areas – especially heavily populated mega-delta regions – will be at a great risk due to sea flooding. In many small island states, water resources will have decrease by the mid-century and sea level rise is expected to exacerbate inundation, storm surge, erosion, and other coastal hazards threatening vital infrastructures.

[53] Intergovernmental Panel on Climate Change. Fourth Assessment Report "Climate Change 2007: Synthesis Report." Geneva: IPCC, November 2007.

3.3 The contribution of earth observation to environmental monitoring and Global Monitoring for Environment and Security (GMES)

Pierre Potin & Josef Aschbacher

This article has two main parts. The first part presents examples of the contribution of Earth observation to environmental monitoring based on scientific results and applications using data primarily from the ERS and ENVISAT missions. The second part presents the Global Monitoring for Environment and Security (GMES) initiative, focusing on the programmatic aspects of the Space Component.[54]

3.3.1. Examples of Earth observation for environmental monitoring[55]

3.3.1.1. Warning signs from space: rising sea level

Climate change has recently become a high priority item on the international political agenda, and observations from space are a strong asset in providing relevant information for the understanding and monitoring of climate change.

In the past years, global warming has accelerated. This is revealed by the faster rate of global sea-level rise in the order of 3.1 mm/year since 1993. The increase is illustrated by Figure 4, which shows measurements derived from ERS-1, ERS-2 and ENVISAT altimeters, and also includes observations from other altimetry sensors. Figure 4 also demonstrates the utmost importance of ensuring the continuity of measurement data sets on a sustainable basis by guaranteeing the availability of the necessary space-based missions in the long term. This is one of the objectives of the GMES programme. Sentinel-3, as part of the GMES Space Component, will ensure the continuity of ERS and ENVISAT altimetry data in polar orbit.

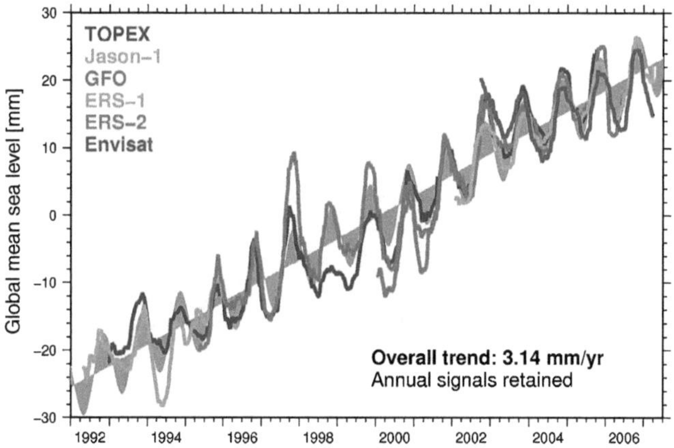

Fig. 4. *Global sea level change from ERS-1, ERS-2 and ENVISAT altimetry (source: Remko Scharroo Altimetrics LLC, Cornish, NH).*

3.3.1.2. Warning signs from space: air pollution and the concentration of greenhouse gases increases

Air pollution has become a global issue. Much of anthropogenic air pollution covers large distances and affects areas far from the emission source. Air pollution affects the health of people, has a damaging effect on the flora and fauna, and causes acid rain. It is related to large-scale fossil fuel combustion and fossil fuel-related activities, but also to biomass burning and changes in land use. The emissions of oxides of nitrogen (NO and NO_2) are an important factor in air pollution. For example, the measurements made by GOME (ERS-2) and SCIAMACHY (ENVISAT) revealed an increase of 50% in the NO_2 concentration over China in the last 8 years.

With SCIAMACHY observations, scientists have recently produced the first time-series showing the global distribution of the most important greenhouse gases – carbon dioxide and methane – that contribute to global warming. Figure 5 shows the average concentration, at a global level, of carbon dioxide during specific periods from 2003 to 2005.

The importance of cutting emissions from 'anthropogenic,' or man-made, gases was highlighted recently when European Union leaders endorsed binding targets for cutting greenhouse gases by at least 20% by 2020, starting from the 1990 levels.

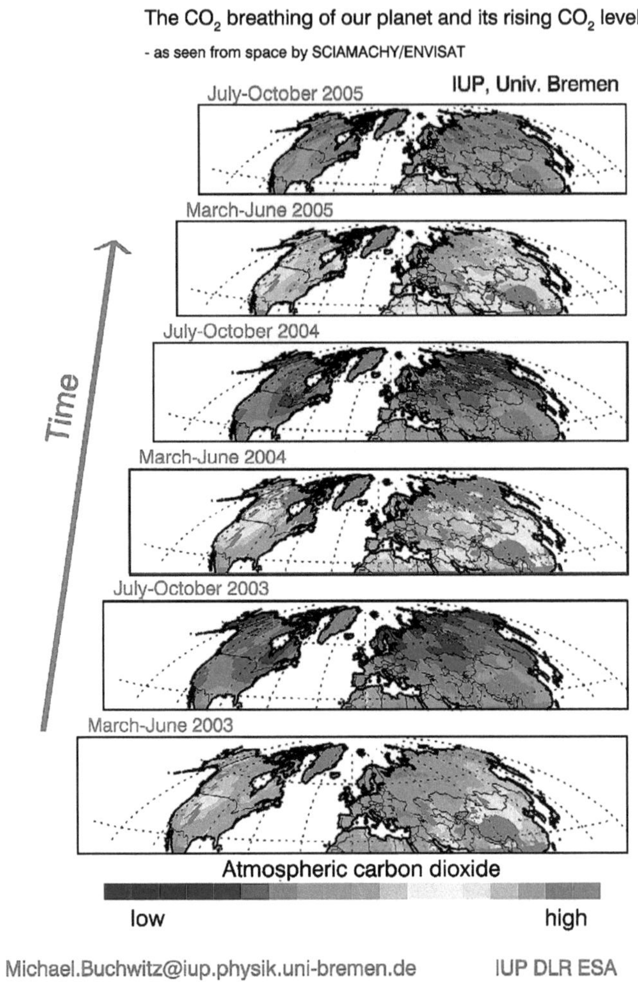

Fig. 5. *Figure 5 shows the average concentration, at a global level, of carbon dioxide during specific periods from 2003 to 2005.*

There are significant gaps in the knowledge about carbon dioxide sources such as fires, volcanic activity and the respiration of living organisms, and about its natural sinks such as the land and ocean. By improving the understanding all of the parameters involved in the carbon cycle, scientists can better predict climate change as well as better monitor international treaties that are aimed at reducing greenhouse gas emissions, such as the Kyoto Protocol which addresses the reduction of six greenhouse gases including methane and carbon dioxide.

3.3.1.3. Emerging Earth observation applications: long swell tracking

During the ENVISAT Symposium held in Montreux, Switzerland, from 23 to 27 April 2007, Chapron and Collard presented for the first time a demonstration of swell tracking across the Pacific Ocean over a period of 12 days using the ENVISAT ASAR Wave Mode (see the snapshot of a time-series in Figure 6).

According to Bertrand Chapron from the French Research Institute for Exploitation of the Sea, "swells are still surprise factors, which can unfortunately be deadly. ... The SAR Wave Mode product allows us to locate and systematically track swells globally. In the near future, we anticipate using SAR wave data to predict their arrival time and intensity".

Waves of all wavelengths and travelling in several directions are generated by storms. Upon leaving the storm, they disperse and those with the longest wavelengths travel the fastest. A larger wave period correlates to a more extreme wind event. Approaching the coastline, the wave system slows down and individual waves increase easily to at least two times the mean average of their initial wave height. For instance, a 5-m significant wave height system can hit the coast with a height of 10 m. As part of GMES, the Sentinel-1 SAR instrument will have a

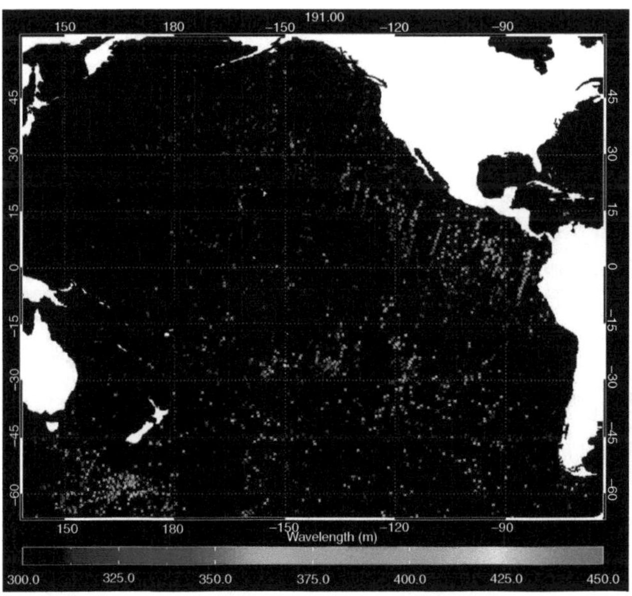

Fig. 6. *ENVISAT ASAR monitors a long swell propagating across the Pacific Ocean (source: B. Chapron, IFREMER, and F. Collard, BOOST Technologies, Brest, France).*[57]

dedicated wave mode allowing the near real-time tracking and forecasting of swells for European users.

3.3.1.4. Understanding global processes and climate change: global terrestrial carbon and vegetation studies

The Leaf Area Index (LAI), one of the Global Climate Observing System (GCOS) Essential Climate Variables (ECV) for land, is an intermediate variable quantifying plant canopy structure which is used in dynamic global vegetation and climate models for the calculation of exchanges of terrestrial energy, carbon and water cycle processes, and the biogeochemistry of vegetation. LAI is critical for most ecosystem productivity models, carbon budget models, and models of global climate, hydrology, and biogeochemistry.

The Globcarbon project as part of the ESA Data User Element activities has the objective of developing a service quasi-independent of the original Earth observation source. The service focuses on the measurement of LAI, FAPAR (Fraction of Absorbed Photosynthetically Active Radiation),[58] burned areas and vegetation growth cycles, and covers 10 complete years (from 1998 to 2007). It uses data from VEGETATION, ATSR-2, AATSR and MERIS. Figure 7 illustrates the average FAPAR indicator for the months of May 2004 and May 2005 over Europe.

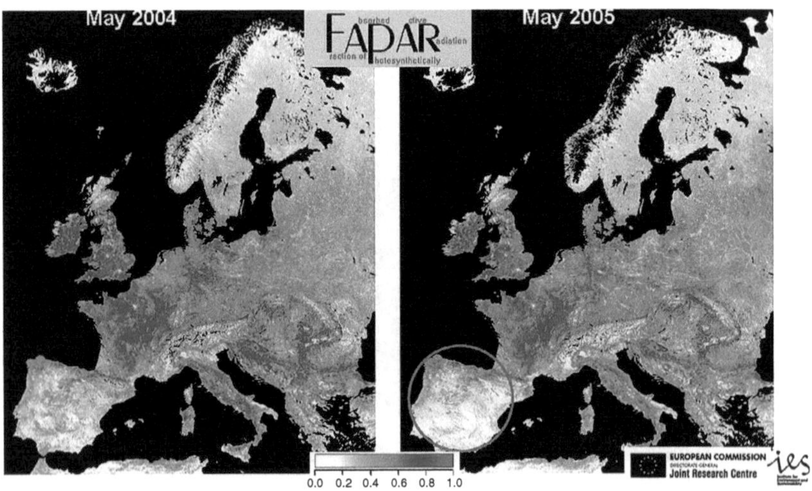

Fig. 7. *FAPAR indicator (Courtesy of Joint Research Centre – European Commission).*

3.3.2. The GMES programme

3.3.2.1. GMES – political and strategic goals

GMES is a European Union-led initiative organised in partnership with ESA to combine ground and space-based observations for developing an integrated environmental monitoring capability. GMES aims at developing operational services following the example of meteorology, but for other domains such as ocean and sea ice monitoring, air quality monitoring, land monitoring, emergency management, security, etc.

Policies addressing the area of environment and security are currently high on the European agenda. Society and politicians need operational information services to effectively manage our planet's environment, understand and mitigate the effects of climate change, and ensure civil security for Europe's citizens. GMES will provide accurate, up-to-date and globally available information on an operational basis to European, national, regional and local entities.

GMES brings to fruition years of research activities concerned with the observation and understanding of Earth system dynamics. The progressive implementation of GMES is made possible by the activities and investments of the European Union and ESA member states. These and other public and private contributions are jointly supported by the EC and ESA.

The Intergovernmental Panel on Climate Change concluded that by the end of this century, global temperatures will increase by 1.8–4°C, and sea levels will rise by 28–43 cm. Heat waves and tropical storms are likely to intensify and Arctic summer sea ice will disappear in the second half of the century.

Climate change will most affect the developing countries. They are already under stress from a steeply growing population which, globally, will increase from 6 bn today to 9 bn by 2050.[60] Shortages of food and clean water will lead to conflicts and the large-scale emigration to wealthier regions such as Europe and North America.

Environment and security are closely related topics and will become major political issues in the coming decades. The command of information has geo-strategical implications. The political mandate for GMES was expressed at the 2001 Gothenburg EU Summit stating the need to "achieve by 2008 an operational and autonomous European capacity for global monitoring for environment and security". This has been transformed into an Action Plan 2004–2008[61] which was subsequently endorsed by the European Parliament.[62] Furthermore, the second Space Council – made up of the EU Competitiveness Council and the ESA Council – stated that GMES will be the second flagship of the European Space Policy after Galileo.

The use of space infrastructure will be a critical component of GMES, assuring autonomous, up-to-date and globally comparable information that will be made available to European and national policy-makers in the domains of environment and security. A secondary objective of GMES is to provide a coherent European contribution to the international efforts such as the Global Earth Observation System of Systems, GEOSS, which was established in 2005 at an international Ministerial Summit. GEOSS has been modelled largely on GMES as regards its overall objectives, but it relies totally on participants' contributions. GMES will be one of the most significant contributors to GEOSS.

3.3.2.2. GMES – a European project in the making

GMES was initiated in 1998 by the "Baveno Manifesto" which states the need for a global satellite-based monitoring system for Europe. GMES gained political momentum in 2000 when it first appeared on the agenda of several subsequent EU Presidencies. In 2001, both ESA and the EC obtained approval for approximately 100 million euros each for initiating GMES services and building a sustainable user community, hence preparing the foundation for GMES.

GMES comprises four main elements that are inter-dependent: (i) services, (ii) space observations, (iii) in situ observations, and (iv) data management, integration and modelling. The latter is of particular importance, as it allows forecasts of parameters critical for the environment and security.

In 2005, the EC assumed the political leadership of GMES and took responsibility for the development of GMES services, including the definition and setup of a sustainable governance structure. Based on the European Space Policy, and for GMES in particular, the EU will use its full potential for identifying and bringing together user needs while mobilising political will in support of these and wider policy objectives. The EU will ensure the availability and continuity of the operational services that support its policies. It will contribute to the development, deployment, and operation of the corresponding European space infrastructure while making maximum use of the existing and planned assets available to Europe, including those of EUMETSAT.

ESA has been made responsible for the implementation of the GMES Space Component. This role comprises two basic functions and assigns responsibility to ESA for:

- The end-to-end definition and implementation of the GMES Space Component as a contribution to the EC-led GMES initiative. Some of the elements are provided by national entities or EUMETSAT;

- The development of space and ground infrastructure through ESA programmes, complementing national/EUMETSAT contributions.

The elements developed through ESA programmes, i.e. the GMES Sentinels and the corresponding ground segment and operations, are complementary to the member state contributions. They have been defined by a gap analysis. In addition to the dedicated infrastructure developed by ESA, the contributions of member states are crucial for GMES. The contributions are mostly in the form of access to data from national and EUMETSAT missions, as listed below.

3.3.2.3. GMES services

The ESA and the EC started developing the GMES information services as early as 2001. Under the ESA GMES Service Element of the Earth Watch programme approved in November 2001, 10 service portfolios were developed. Each of these portfolios addresses user needs in specific sectors of environmental and security policy. They address domains such as polar monitoring, forest monitoring, marine and coastal monitoring, flood, fire and geo-hazard risk assessment, air quality monitoring and forecasting, land cover mapping and urban development monitoring, as well as information services to support humanitarian aid or development and food security activities. At the same time, the EC launched a series of large, integrated projects within the Sixth Framework Programme to address the underlying research issues. Together, these activities involved more than 400 organisations from 35 countries as end users of GMES. This enabled the ESA and the EC to identify the key requirements for the future GMES space infrastructure, and led to specifications for the Sentinels and access to national missions.

Building on these achievements, the EC organised major consultations with user organisations in Europe and is now dedicating significant resources under the Seventh Framework Programme to the development of the three fast track services: the Land Monitoring Core Service, the Marine Core Service, and the Emergency Response Core Service. These fast track services will enter their pre-operational stage in 2008. Two other "pilot services" are planned: the GMES Atmospheric Service and a service dedicated to security. The implementation follows a layered service model, with a set of pan-European or global core services being established first. Later, more specialised downstream services that can be derived from the core services, for example at the national, regional, or local level, will be implemented.

3.3.2.4. Responding to user requirements

The content of the GMES Space Component is a response to user requirements. The definition of the GMES Sentinels, which were approved at the Berlin ESA Ministerial Council of December 2005, was based on the results of a documented exercise. This exercise analysed user requirements based on policy needs, derived space observation requirements and compared them with available and planned missions from EU/ESA member states and EUMETSAT and, based on this gap analysis, defined the ESA-developed Sentinel missions as a complement to the said missions. This is illustrated in Figure 8.

The EC has been closely involved in defining the details of the Sentinels. The EC has organised the assessment of the Mission Requirement Documents (MRDs) of Sentinels-1, -2 and -3 through the GMES Fast Track Implementation Groups on emergency management, marine and land monitoring. The respective recommendations have led to design changes of the Sentinel missions in a few cases. Based on these reports, the EC has prepared a synthesis report on space infrastructure needs that endorses the design of the ESA-developed Sentinel-1, -2 and -3 missions. The report also states that two spacecraft of each Sentinel (-1, -2 and -3) are needed for satisfying the observation requirements in terms of coverage and repeat cycle.

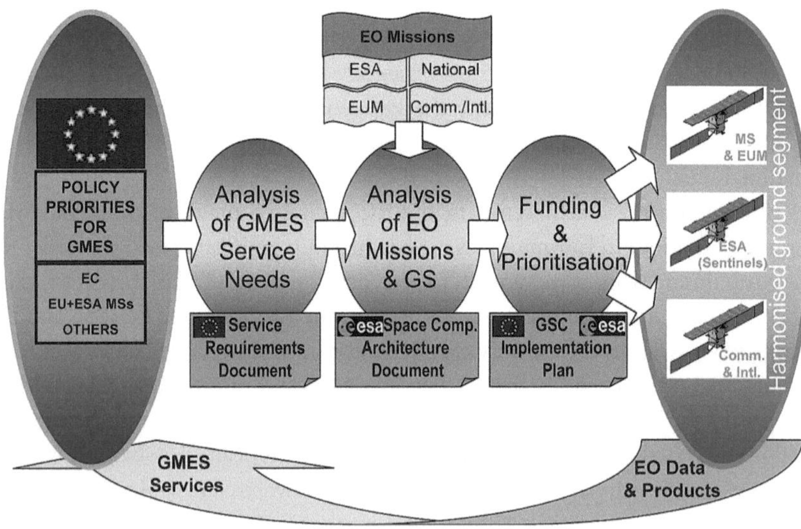

Fig. 8. *The process leading to the GMES satellites.*

3.3.2.5. National and multi-national contributions

The GMES Space Component comprises both shared infrastructure provided by national, European, and other contributors, and dedicated infrastructure (Sentinels) developed by ESA.

Today, a number of national, multi-national (including EUMETSAT) and commercial EO missions are available in Europe. They primarily serve their respective operator's strategic priorities, but may also be of interest to GMES. Currently, the following Earth observation missions are identified as potential contributors (in alphabetic order): Altika, Cosmo-SkyMed, Cryosat-2, DMC, EnMap, ENVISAT, ERS, Jason, Meteosat, MetOp, Pleiades, PROBA, Radarsat, RapidEye, SeoSat, Spot, Tandem-X, TerraSar-X, and TopSat. In order to secure the sustainability of those national and multi-national missions that are required for GMES (based on a formalised user endorsement process), a high-level coordination group has been set-up by ESA and the EC involving all stakeholders.

3.3.2.6. Sentinel missions

Based on the gap analysis, five observation capacities have been identified that need to be developed for GMES in addition to national and multi-national missions. These dedicated GMES missions, called the Sentinels, are defined as follows:

- Sentinel 1 – High-resolution SAR imaging
- Sentinel 2 – High-resolution multispectral imaging
- Sentinel 3 – Global ocean and land monitoring
- Sentinel 4 – Geostationary atmospheric composition monitoring
- Sentinel 5 – Low-orbit atmospheric composition monitoring.

The Sentinels have the following characteristics.

The Sentinel-1 mission carries a SAR in a precise dawn–dusk sun-synchronous orbit at a 700-km altitude with an exact repeat of 12 days in support of multi-pass interferometry. With the SAR swath of approximately 240 km, a 12-day quasi-global coverage is ensured. The ground resolution of approximately 5 m exceeds that of ERS and ENVISAT in imaging mode. The satellite, weighing about 2.1 tonnes, is adapted to the large SAR instrument which features a phased-array antenna with 5 kW total radiated power. A duty cycle of 20% is expected. The satellite is designed for launch on Soyuz and has a design lifetime of 7 years (Figure 9).

Fig. 9. *Artist's impression of Sentinel-1 (Source: ESA).*

The Sentinel-2 mission carries a medium to high-resolution push-broom multi-spectral imager operating in VNIR (Visible and Near-InfraRed) and SWIR (Short-Wave InfraRed) in an ENVISAT-like orbit. The ground resolution is 10, 20, and 60 m (depending on the channel). The swath width of the multi-spectral imager is approximately 280 km, which ensures the systematic acquisition of all land surfaces every 10 days. The satellite, weighing about 850 kg, fits on a small launcher (Vega) and has a 7-year design lifetime (Figure 10).

Fig. 10. *Artist's impression of Sentinel-2 (Source: ESA).*

Fig. 11. *Artist's impression of Sentinel-3 (Source: ESA).*

The Sentinel-3 mission carries a Cryosat-derived microwave altimeter (including a microwave radiometer and a precise orbit determination device) and two imagers for ocean/land colour observations (MERIS-like) and for sea/land surface temperature observations (AATSR-like) into an ENVISAT-like orbit. The imager also ensures the continuity of the VEGETATION mission. The satellite, weighing approximately 1.3 tonnes, supports the 100% instrument duty cycle and is launched with a small launcher (Vega). The design lifetime is 7 years (Figure 11).

Sentinels-4 and -5 are atmospheric chemistry missions, details of which are currently being defined in pre-Phase A studies. The baseline assumption is that they are flown as instruments on Meteosat Third Generation (MTG) with a launch around 2017 (Sentinel-4) and on post-EUMETSAT Polar System (Post-EPS) with a launch around 2019 (Sentinel-5). The utilisation of these platforms, if confirmed in the above configuration, will be closely coordinated with EUMETSAT which also operates the MTG and post-EPS spacecraft.

Two spacecraft of Sentinel-1, -2 and -3 are needed in orbit to fully meet the coverage and observation frequency requirements. The long-term continuity of Earth observation data is a pre-requisite for GMES services. It is therefore assumed that the baseline technology for the space segment will remain stable for a longer period, e.g. 15 years per satellite generation, which is similar to other operational EO programmes such as Meteosat, Spot, Landsat, and NOAA/AVHRR. This assumption does not exclude some evolutionary improvements to the space segment in respect of in-orbit performance, for example. Likewise, the ground segment will have shorter technology renewal cycles to take account of better computing, telecommunications, software systems, and user interface technologies.

The strategy and phasing of the procurement process will significantly influence the overall GMES Space Component programme's financial profile and costs. The spacecraft deployment schedule, lifetime, overall funding and the use of new or existing technologies are key parameters that have an impact on global costs. For this reason, the choice of technology tends to keep the costs of the overall system low, rather than increasing performance. Technology has to be mature prior to entering the phase of engineering and system development.

3.3.2.7. Ground segment, data access and operations

3.3.2.7.1. Payload data ground segment implementation and evolution

The ground segment is implemented to control and exploit the ESA Sentinels-1, -2 and -3 spacecraft and to manage, plan and monitor the overall GMES Space Component, including contributing missions. The Payload Data Ground Segment is a distributed ground segment that reuses existing facilities, infrastructure, and expertise. Furthermore, the ground segment links and shares already available infrastructure developed for national missions.

The operation and utilisation of Sentinels has a higher degree of automation compared to ERS and ENVISAT, for example, but near real-time requirements and operational reliability will place higher demands on the development of the ground segment.

3.3.2.7.2. Data access and operations

From 2008 onwards, pre-operational support of GMES services will start in the form of co-ordinated and harmonised EO data provided by national, EUMETSAT and other third party missions, all referred to as GMES Contributing Missions. The routine operation covers the overall GMES Space Component, i.e. not only the operation of ESA Sentinel spacecraft, but also the operation of the other GMES missions which in turn contribute to GMES. This includes mission planning across all missions, a distributed user service shared with all partners, a coherent quality assessment and the monitoring and control of the end-to-end data provision for the entire GMES Space Component.

3.3.2.7.3. Socio-economic benefits of GMES

A consortium led by PricewaterhouseCoopers (PWC) has analysed the potential socio-economic benefits of GMES for the period 2006–2030 on the basis of a wide-ranging consultation with stakeholders and authorities in the policy sectors that GMES addresses. The study demonstrated that GMES could entail significant socio-economic benefits as a consequence of future decisions in the field of environment and security policy that will have to be supplied with better, more complete, consistent, timely, and reliable information from GMES. Significant socio-economic benefits were identified and quantified in the following policy areas:

- Europe as a global partner (climate change adaptation, global environment protection, humanitarian response)
- Preservation and management of natural resources (air quality, marine environment, forest ecosystem management, civil protection).

Significant benefits were also identified, but not financially quantified, in the following domains:

- Europe as a global partner (climate change mitigation, development and aid)
- Preservation and management of natural resources (urban and rural policy, agricultural policy, water quality, management of wetlands).

GMES strategic and political benefits were grouped as follows:

- Efficiency gains in the implementation of existing policies: these gains could be realised almost immediately. They represent a monetary value of 100 million euros per annum.
- European policy formulation benefits: since these depend on the future development of policies, these benefits would accrue later, typically with a lag of one decade. PWC estimates the potential magnitude of these benefits to be at least 10 times greater than in the first category.
- Global action benefits: since they depend on new international policy agreements, these benefits would also accrue later. Similarly, the external dependencies and uncertainties of their realisation are greater. However, by virtue of their global scope, they carry the greatest potential benefits. We estimate the benefits to be at least 10 times greater than those in the previous categories.

The potential GMES benefits accumulated over the 25 year period from 2006 to 2030 were estimated as comparable to 0.2% (approximately 80 billion euros) of the EU's current GDP. The study highlighted the fact that the mere availability of information does not in itself translate into a benefit. The GMES benefits will materialise only when the information is used. Therefore, during the implementation phase, major emphasis will be placed on integrating GMES information into future policy and decision-making processes within Europe.

3.3.2.8. Phased funding approach

3.3.2.8.1. Period up to 2013

The funding required for building the GMES Space Component, or GMES In-Orbit Operational Validation (IOV), is obtained in phases. ESA funds are complemented by EC funds.

Several funding steps are planned for the period 2006–2013. On the ESA side, the 2005 Ministerial Council provided 257 million euros for the first phase of the GMES Space Component programme. This was then supplemented by an additional 501 million euros approved in September 2007 for phase 2. At the 2008 Ministerial Council, further funding will be adopted for Segment 2 of the programme.

The EC has earmarked a total of about 700 million euros through its FP7 Space Work Programme for the period 2007–2013. Other, future contributions will be necessary to cover the operational phase of GMES beyond 2013 and to complete the IOV of the current Sentinels (e.g. Sentinels-4 and -5 that will be launched in the 2017–2019 period). All figures are given in 2006 terms.

3.3.2.8.2. Long-term funding considerations

Regarding the space infrastructure, the long-term funding sources have to move from research and development to infrastructure budget lines. The current share of the investments made via ESA member states for building the GMES Space Component (GMES IOV) is expected to decrease, while the EC is expected to increasingly replace its research and development contributions with infrastructure budgets to support operations (including recurrent spacecraft).

The goal should be to follow the example of meteorology where ESA is developing and financing – with an adequate contribution from EUMETSAT – the first prototypes based on the requirements defined by the users that are responsible for funding the recurrent units, operations, and services. Funding by ESA would focus on technological evolution and upgrades as well as on the overall coordination of the GMES Space Component. The development of funding sources as described above would be in line with the aim of making GMES an operational system that provides critical information in support of European policy goals.

Acknowledgements

The authors wish to thank the ESA colleagues involved in the compilation of this article, in particular colleagues from the EO Services and Exploitation Division who provided information on the first part. Moreover, we would like to thank the scientists who provided illustrations and animations included in this article.

---

[54] A limited number of illustrations are provided in this paper. Additional material is available in the slides of the presentation prepared by the European Space Policy Institute and available at the following link and upon request by e-mail to the authors for the complete set of slides, including the animations: http://www.espi.or.at/images/stories/dokumente/conferences/2007/potin.pdf

[55] The examples shown in this section are only a few demonstrations (related to seas/oceans, atmosphere and land) of how Earth observation can contribute to the monitoring of our environment at a global level.

[56] The animation, part of the presentation, covers three years, from 2003 to 2005, with one-month intervals.

[57] This illustration is a snapshot only. The animation, part of the presentation, covers 12 days at 6-h intervals.

[58] FAPAR is an indicator of the state and productivity of vegetation and thus a fundamental surface parameter for environmental studies.

[59] This illustration is a snapshot for the month of January 1999. The animation, part of the presentation, covers four years from 1999 to 2002 at 1-month intervals.

[60] Plan of Implementation of the World Summit on Sustainable Development (WSSD), 26 Aug.–4 Sept. 2002, Johannesburg.

[61] European Commission. Global Monitoring for Environment and Security (GMES): Establishing a GMES Capacity by 2008. Action Plan 2004–2008, COM (2004) 65 final, 3 Feb. 2004.

[62] European Parliament. European Parliament Resolution on the Action Plan for Implementing the European Space Policy, B5-0045/2004, 22 Jan. 2004.

3.4 Environmental threats and space responses

Jérôme Lafeuille

3.4.1. Introduction

In spite of the impressive technological advances achieved, our society remains particularly vulnerable to environmental threats. Some of these threats are linked with severe weather conditions like short-term cyclones or tornadoes, heavy rains and subsequent flooding. Other threats can be linked to meteorological conditions occurring over a longer period of time through a cumulative process, such as weeks or months of heat waves or drought, with potential consequences ranging from decreasing air quality and public health hazards to the risk of wild fires or the discontinuity of the food supply, depending on the natural and socio-economic context. On a different note, the growth of the global population and economic activity is a direct cause of an increasing stress on natural resources, which obviously threatens the current state of our natural environment. As a further consequence, it is now established that global warming is primarily the result of an increased concentration of greenhouse gases in the atmosphere and that this will have a significant impact at the regional level, posing new specific threats differing from region to region. In addition, the expanded use of high-technology automated services relying on radio-communications and satellite capabilities creates a new exposure to space weather phenomena and issues of the sustainability of outer space, which should be paid due attention. Space-based capabilities, particularly Earth observation, have an important role to play in providing means for the monitoring of these threats and for assisting in their mitigation.

3.4.2. Aggravating environmental threats

While the relationship between mankind and the global environment has always involved some risk, the nature and degree of that risk has changed as a result of the growing world population and the environmental impact of human activity. Among the changes that can be observed, three particular aspects are highlighted here, as they appear to be aggravating factors of environmental threats, thus raising long-term sustainability issues: the increased societal exposure to severe weather events; the increased stress on our environment; and the effects of climate change. These factors are considered in more detail below.

3.4.2.1. Societal vulnerability to severe weather

Every month, new evidence appears of the dramatic impact of severe weather events in terms of material damages and human casualties around the globe. As an example, in November 2007, more than 3,500 people died when Bangladesh was hit by the tropical cyclone Sidr. In the same month, heavy rains generated massive floods affecting more than 2,000 houses in Indonesia. Moreover, 2,000 people had to be evacuated in Malawi and 20,000 sought shelter in Mexico. The following month, 20,000 people were displaced by monsoon rain floods in Sri Lanka, and 460,000 people in the U.S. were left without power for hours following an ice storm. In late January 2008, more than one million square kilometres covering over 15 Chinese provinces were affected by exceptional snow and freezing rain precipitation events that temporarily blocked the rail and aircraft circulation of millions of people and motivated the launch of an emergency plan. In February of that year, tropical cyclone Ivan killed 83 people and left 200,000 homeless in Madagascar, followed in March by tropical cyclone Jowke that hit Mozambique. Furthermore, hurricane Katrina, which devastated the area of New Orleans, USA, in August 2005, reminded the world that even the most technologically advanced countries were exposed to dramatic casualties from weather events.

According to the OFDA/CRED International Disaster Database, the number of natural disasters reported is steadily increasing. In particular the economic losses

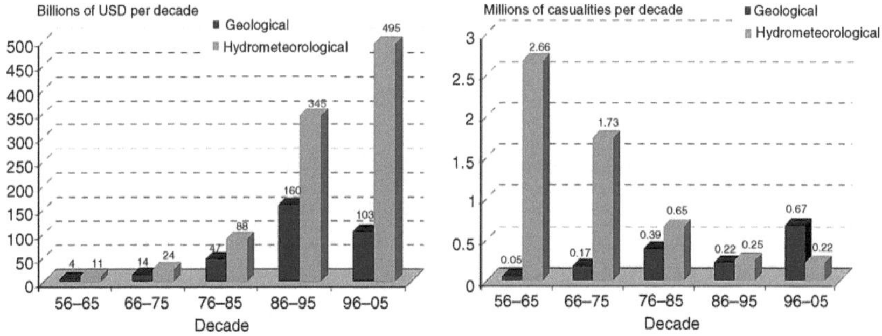

Fig. 12. *Evolution of economic losses (left panel) and human casualties (right panel) due to geological and hydro-meteorological disasters over the past five decades (source: EM-DAT: The OFDA/CRED International Disaster Database; charts by: J.-B. Migraine, WMO Disaster Risk Reduction Programme).*

resulting from hydro-meteorological events have been constantly and sharply rising over the past five decades, from 11 to 495 billion dollars per decade, as shown in Figure 12. Over the same time span, however, human casualties have constantly and significantly decreased, from 2.48 to 0.22 million lives per decade. Although every effort should be made to further reduce these figures, they suggest that progress on disaster preparedness and mitigation activities has enabled societies to efficiently cope with the increased occurrence of disasters and the growing exposure of populations and infrastructures to them.

Besides the effects of global warming on the frequency and intensity of such severe events as discussed below, there is evidence that the exposure of our societies to severe weather is increasing as a result of population growth, the development of urban concentrations and infrastructures, and increased activity and traffic.

3.4.2.2. Increased environmental stress

The growth of the world population and the associated development of economic activity are triggering the accelerated exploitation of natural resources such as fresh water, ocean livestock, forest and agricultural land, while also having a direct bearing on air, soil and water pollution. This development is threatening the long-term availability of vital resources, at least for some parts of the globe. Moreover, it has uncontrolled effects on our climate.

3.4.2.3. Effects of climate change

The Nobel Prize-winning WMO – UNEP Intergovernmental Panel on Climate Change (IPCC) has acknowledged that the global warming of the climate system is unequivocal. It has established that most of the warming is very likely due to positive radiative forcing induced by the anthropogenic increase of the greenhouse gas concentration in the atmosphere over the last 50 years, namely carbon dioxide released by the use of fossil fuel and changes in land use, as well as methane and nitrous oxide released by agricultural practices. In its Fourth Assessment Report, the IPCC states that at the continental, regional and ocean basin level, numerous long-term changes in climate have been observed. These include changes in arctic temperatures and ice, widespread changes in precipitation amounts, ocean salinity, wind patterns and aspects of extreme weather including droughts, heavy precipitation, heat waves and the intensity of tropical cyclones. It is anticipated that continued greenhouse gas emissions at or above the current rates would cause further warming and would induce many changes in the global climate system,

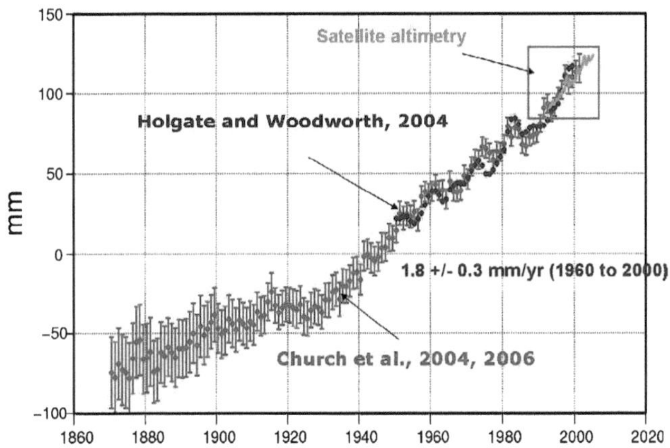

Fig. 13. *Global sea level rise from 1860 until today (source: Stan Wilson, NOAA, and François Parisot, Eumetsat).*

which would very likely be larger in the 21st century than those observed in the 20th century. The exact prediction of the nature and importance of regional changes is a matter of investigation and includes uncertainty margins, mainly because the socio-economic growth scenario cannot be accurately predicted, but there is evidence that this change will generate a range of phenomena such as: sea level rise with induced risk of coastal area flooding, more frequent drought and flooding in some regions, thawing of permafrost and reduction of the arctic polar cap, change of temperature and precipitation patterns that will cause the extinction of some vegetal and animal species in some regions and their prospering in other regions, and changes in the propagation of infectious diseases. Droughts have become more common, especially in the tropics and sub-tropics. At the same time, there has been an increase in the number of heavy precipitation events. Within the foreseeable future, millions of more people will be exposed to costal area flooding each year, while hundreds of millions of people will suffer from water shortages (Figure 13).

3.4.2.4. Technology vulnerability to the outer space situation

Fantastic progress continues to be made in advanced technologies, which has led to an extensive reliance on satellites for telecommunications and for precise localisation services such as the Global Positioning System (GPS). It is expected that, in the future, more and more day-to-day operational services will exchange data via satellites and use GPS-based remote monitoring, for instance to support car and truck traffic management, aircraft and ship navigation, and farming operations.

With the increasing reliance on these technologies, there is an emerging awareness that their reliability is affected by solar events such as solar eruptions, coronal mass ejections or geomagnetic storms, which are referred to as "space weather". Space weather is known to be a major cause of the disruption and aging of in-orbit systems, and it is also a primary source of the temporary inaccuracy of GPS localisation. Space weather events also affect wide-scale electricity distribution networks on the Earth's surface. Furthermore, they raise safety issues for long-distance aeronautical traffic over polar regions, which has experienced a boom in the past decade.

Generally speaking, in the long term, the expanded reliance of operational services on space-based capabilities will require more attention to some issues of outer space sustainability. This should include the proper management of the electromagnetic frequency spectrum to avoid interferences among emitting sources and to protect the natural emission frequency bands of the atmosphere and the Earth's surface, as these bands are unique tools for monitoring the planet through passive radiometry.

3.4.3. Space for mitigating threats

Space-borne capabilities offer essential risk management solutions for mitigating the environmental threats discussed above, in particular through Earth observation satellites. Other space-based capabilities such as data collection, telecommunications and positioning also offer important support for emergency operations.

3.4.3.1. Different steps of risk management

The Hyogo Framework for Action 2005–2015 defines an overall Disaster Risk Management Framework that includes five major components: risk identification and assessment, risk reduction (prevention and mitigation), and risk transfer, supported by information and knowledge sharing, and organised within a governance and organisational coordination and cooperation dimension.

From a chronological point of view, activities can be distinguished as either being conducted *before* a risk materialises (i.e. risk identification and mapping, emergency planning and preparedness), *during* the occurrence of a crisis (early warning, detecting the hazardous event, monitoring and forecasting its evolution, managing the emergency response and potential rescue operations), or *after* the crisis has occurred (final damage assessment, reconstruction, improvement of mitigation strategies).

Earth observation plays a key role in all these stages for environment-related disasters, whether they be of hydro-meteorological, geological or even artificial origin like a severe marine, atmospheric or radio-active pollution crisis, as illustrated by some examples below.

3.4.3.2. A space-based Global Observing System for monitoring, forecasting and warning

High-resolution Earth observation from space is the main support for objective risk mapping before the event and of damage assessment afterwards. High or medium resolution imagery is essential as input for modelling hazardous phenomena. Near-real time meteorological and environmental observation supports early warning and the detection and forecasting of either the event itself or of the hydro-meteorological conditions affecting its evolution or propagation; it supports decision-making during emergency operations.

Most environment-related disasters have indeed either a hydro-meteorological origin, which is obvious in the case of storms, floods, or drought, or are affected in their evolution by hydro-meteorological factors such as, for instance, volcanic eruptions releasing huge ash and SO_2 clouds. Moreover, weather forecasting is an important input in the conduct of emergency operations, as when the mountainous regions of Pakistan were hit by an earthquake while the survival of the population and the shipment of rescue materials were made difficult by heavy snowfall.

The space-based component of the WMO Global Observing System includes operational geostationary and polar-orbiting components, as well as a range of contributing Research and Development (R&D) satellite missions that will progressively evolve into new operational capabilities. In January 2008, nine geostationary and seven polar-orbiting satellites were considered in operational status, a dozen of R&D satellites were contributing, and there were plans for launching up to 17 new satellites in 2008, either as a replacement and enhancement of continuing operational missions, or in order to demonstrate new technologies and obtain research data. The operational geostationary system is a globally coordinated constellation of six satellites with visible and infrared imagers, some sounding capabilities and in-orbit redundancy, which is particularly suited for permanent monitoring, rapidly evolving weather observation and event detection. The core polar-orbiting constellation includes sun-synchronous satellites operated on regularly distributed orbital planes with respect to the sun's direction, i.e. at mid-morning, early afternoon and early-morning or dawn–dusk local solar time, with imagery instruments and, on most of the orbital planes, infrared and microwave sounding. Emphasis is put on the timely collection and delivery of data, whereby geostationary data are typically refreshed every 15–30 min and available to the users in processed form as early as possible within this time window; for polar-orbiting data, only regional data can be available in near-real time, since one has to wait for an orbital revolution (about 101 min) to get hold of the full orbit data that are stored aboard the satellite.

3.4.3.3. From core weather observation to environment monitoring

The scope of the WMO Global Observing System has changed since climate monitoring was recognised as a core objective of the system besides operational weather forecasting. This entails additional long-term continuity constraints, requirements for inter-comparability and the calibration of sensors, as well as observation needs for additional parameters such as Earth radiation budget, atmospheric chemical composition and aerosol content, sea surface topography, ocean colour as a tracker for carbon dioxide absorbing phytoplankton, vegetation type and extent, and terrestrial albedo. Thus, the space-based observing system shall be enhanced in the coming decades by satellite missions addressing these parameters on an operational basis through a range of sensors including radar altimeters, polarised microwave radiometers, high spectral resolution infra-red spectrometers, and specific imagers with either particular spectral resolution or high geometrical resolution. Through this significant development which has

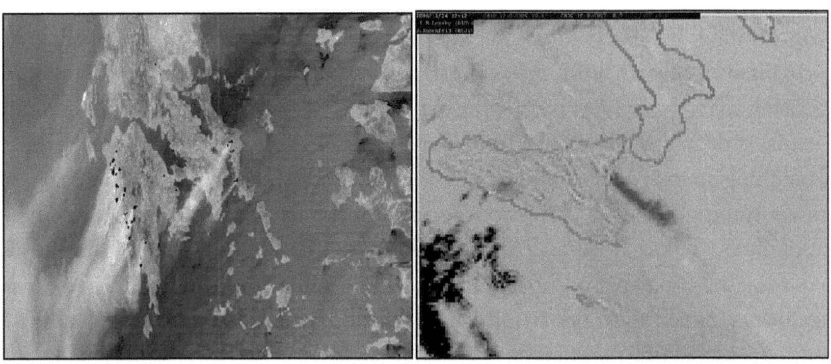

Fig. 14. *Monitoring various hazardous environmental events by meteorological or environmental satellites. Left: Hot spots from wild fires in Greece, as seen by MODIS instrument aboard NASA's Aqua satellite on 26 August 2007 at 11:10 UTC; red–green–blue colour composite based on 3.7, 0.5 and 0.6 μm channels. Right: Infrared signature of a volcanic ash cloud released by Etna as detected by Meteosat-8 on 24 November 2006 at 12:12 UTC; red–green–blue colour composite based on 10.8, 8.7 and 12 μm channels (source: Dany Rosenfeld, University of Jerusalem, Israel, by courtesy of Eumetsat).*

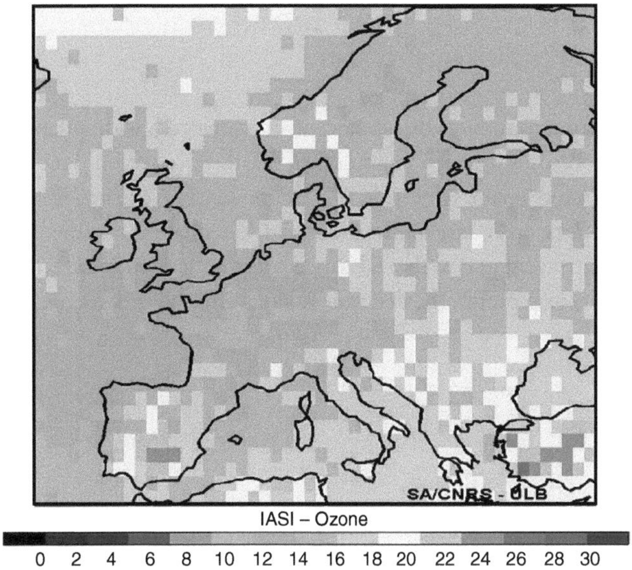

Fig. 15. *Ozone pollution peaks detected over Europe on 22–26 July 2007 by processing observation data from the Infrared Atmospheric Sounding Interferometer (IASI) aboard the Eumetsat Metop satellite (source: Cathy Clerbaux, CNRS, France, by courtesy of Eumetsat).*

already started, the space-based Global Observing System will progressively become a comprehensive platform for monitoring a number of environmental features beyond pure meteorological phenomena. Through this development, the space-based Global Observing System is becoming a major component of the Group on Earth Observation's (GEO) Global Earth Observation System of Systems (GEOSS), directly or indirectly addressing a number of its Societal Benefit Areas (SBA) (Figure 14).

Tropospheric air quality, as a combined result of climate change and the accompanying more frequent and extended heat waves in certain regions on the one hand, and of urban concentration and industrial growth with the resulting peak levels of ozone concentration (as observed in various parts of Europe during recent summer seasons) on the other hand, has become an important concern (Figure 15).

3.4.4. Challenges

Space-borne systems offer highly relevant capabilities assisting in facing the current and future environmental threats. However, making the best use of these capabilities does not go without a few challenges that are highlighted below.

3.4.4.1. Integrating space-based observation in a decision chain

Experience from numerous disaster events has created a widespread awareness of the importance of a proper integration of Earth observation data interpretation into the decision chain. The role of remote sensing experts is to ensure that observation data are consistently processed, adequately qualified and properly interpreted. Efficient and well-established communication between the originator of the data, its interpreter, the decision-makers and the end-users is critical to the overall process.

3.4.4.2. Enhancing the space-based Global Observing System

The evolution from a purely meteorology-focussed system to a comprehensive system encompassing climate and environment monitoring is a wide-ranging integration process which can only be achieved through internationally coordinated efforts. This is to ensure that the addition of individual plans will fit into a comprehensive system with no gaps in essential observations, and that the data provided by different satellite systems operated by different countries will be exchanged and will be calibrated against common references making them intercomparable and seamlessly mergeable into consolidated datasets.

Different mechanisms are in place to support this coordination. First of all, WMO supports the definition of observing requirements by representative users of different application areas and maintains a consolidated depository of these requirements in terms of geophysical parameter, horizontal, spectral and temporal resolution and accuracy. WMO also supports the regular critical review of the available and planned observing capabilities to fulfil these requirements, which results in a gap analysis and proposals for optimising the design of the Global Observing System. WMO is also an intergovernmental forum where members make high-level voluntary commitments of contributing to the global system, which over the past 60 years has enabled the successful implementation and operation of worldwide interoperable systems for the meteorological community. Implementation plans to respond to space-related WMO needs are discussed within the Coordination Group for Meteorological Satellites (CGMS), which for the past 35 years has been the main forum for technical coordination among satellite operators contributing to WMO programmes. The initial focus was on operational missions in support of weather forecasting, enabling the robust geostationary and polar-orbiting constellations that we now rely on. In addition, the Committee on Earth Observation Satellites (CEOS), which was established 22 years ago with a larger representation of Research and Development (R&D) space agencies, has adopted an action plan to provide a coordinated response to climate monitoring needs and, in a wider perspective, to GEOSS; this has led CEOS to considering different "constellations", either virtual or real, along which CEOS member space agencies can focus their planning and development efforts. WMO, CGMS and CEOS are working together and building on each other's efforts, thus combining the input of user communities, operational agencies, R&D agencies and governments in order to implement and enhance the Global Observing System.

3.4.4.3. Ensuring long-term sustainability

A major challenge today is to ensure that the observation needs for climate and environment monitoring will be fulfilled on a long-term basis, because many of the needs have only been addressed thus far in the context of R&D programmes with no objective of long-term continuity. This has been due to the scientific research needed on climate processes and the experimental nature of the sensors used for collecting relevant data in the past decades. Experimental and scientific missions remain essential objectives nowadays, but climate monitoring has reached a stage where operational continuity is needed in collecting and processing observation data for what the Global Climate Observing System (GCOS) has defined as the "Essential Climate Variables". The transition from R&D to an operational status is not only a technological evolution, but also implies the recognition of a new operational need for our society. This has financial and organisational implications, since the past sharing of responsibilities between operational and R&D space agencies might not be sufficient; national or international authorities must identify or establish relevant capabilities to face this challenge and, for instance, consider providing space agencies (either operational, R&D or through partnerships of the two) with funding and a mandate to support climate monitoring programmes on a long-term basis.

3.4.5. Conclusion

Mankind is exposed to multiple environmental threats and many of these threats are expected to worsen in the coming decades. Space-based capabilities, and in particular Earth observation, play a unique role in assisting in the management of these threats. This refers both to planning and evaluation processes and to actual monitoring and real-time support in critical situations, as in the case of hydro-meteorology related events.

Thus, there are many reasons to enhance the space-based Global Observation System towards a comprehensive monitoring system for climate and environment, which is the current goal of the WMO in partnership with GCOS, CGMS and CEOS, and a major contribution of GEO to the GEOSS. At the European level, the Global Monitoring for Environment and Security (GMES) provides a framework for supporting such initiatives.

Action requires resources and since decisions are normally taken in the light of cost-benefit aspects, it should be borne in mind that the main societal benefit of mitigating a threat is the minimisation of economic losses and casualties that would be incurred if no action were taken to improve our preparedness for facing the future.

CHAPTER 4

RESOURCES

4.1 Overview

The context: not only the developing world, but also Europe has begun to deal with the scarcity of natural resources, including water. Because food and water are considered to be elements of the universal human rights, every human being should have adequate access to them. Due to several reasons, including climate change, the land suitable for agricultural uses has been decreasing, resulting in a need for an increased trade in food as well as better for farming strategies.

The contribution of space: space can be of extreme value in assuring food security in the world through applications such as remote sensing. The latest space technologies can contribute to the effectiveness of humanitarian responses in areas where there is no adequate access to food and water. The utilisation of baseline mapping, crop modelling, agricultural monitoring and disaster management as well as precision farming can all greatly benefit the agricultural sector and hence have a positive impact on food security.

The papers: the three papers of this section provide a broad overview of the natural resources on Earth and the potential threats to them. Food security and water management appear to be the dominant themes.

In "Food Security and Climate Change: Remote Sensing and Sustainable Agricultural Development", Mahendra M. Shah from the International Institute of Applied Systems Analysis (IIASA) argues that climate change is a global issue that will have long-term implications and involves complex interaction between demographic, climatic, environmental, economic, health, political, institutional, social and technological processes. Food security is a growing problem that needs to be addressed immediately, for which space technology can offer feasible solutions.

In "The Use of Space Technologies for Effective and Sustainable Responses to Food Insecurity Threats in Africa", Menghestab Haile from the World Food Programme (WFP) gives an overview of food security issues (mainly in Africa). He then goes into discussing the potential contributions of space technologies to the reduction of poverty in the world. In detail, he describes the usefulness and relevance of remote sensing, satellite-supported communications, baseline mapping, and environmental monitoring.

In their paper entitled "Threats, Risks and Sustainability: Answers from Space Earth Observation Data for the Management of Renewable Resources", Rupert

Haydn et al. from GAF AG discuss the use of space-borne Earth Observation (EO) data for to the assessment of renewable resources. The article is based on case studies which represent contractual activities relating to water, forests and crops carried out by GAF on behalf of the European Space Agency (ESA), the European Union (EU), the World Bank (WB), and others.

4.2 Food security and climate change: remote sensing and sustainable agriculture development

Mahendra M. Shah

4.2.1. Food security and hunger

Food is universally recognised as a fundamental human right and at the global level, there is enough food to meet everyone's need. Yet a fifth of the world's population continues to be chronically undernourished. Every year, about 15 million people die from hunger and hunger-related causes,[63] and over 200 million suffer health consequences due to deficiencies such as a lack of proteins, micronutrients, and essential amino acids.[64] Balanced nutrition is the foundation of good health and healthy people are less susceptible to many infections and diseases.

In 1974, the world's political leaders set the goal of eradicating hunger within a decade. A quarter of a century later, at the Millennium Summit[65] in 2000 and the World Food Summit in 2002, the world's leaders endorsed the less ambitious goal of reducing hunger by half by 2015, even though the rate of progress over the 10 preceding years indicated that it would take more than 60 years to reach this target. Political goals have a role and relevance, but there is a limit to the hope and trust of hundreds of millions who spend a lifetime of debilitating hunger.

Beyond the overarching problem of world hunger, today there is also an emerging problem of "over-consumption" that results in obesity and related health disorders such as diabetes and cardio vascular diseases. Already more than 800 million people worldwide are estimated to be obese. The next world food crisis will no doubt be one of human health, be it due to too little food or too much unhealthy food that will affect people differentially in all countries, developing and developed.

Food agriculture must be given the highest political commitment and attention because it is the dominant user of environmental and natural resources, and has the greatest impact on the sustainability of ecosystems and their services. The world food system also directly and indirectly accounts for the employment and livelihoods of as much as half of the world's working adult population. However, the agricultural sector is often not given the attention and commitment that it deserves, especially in developing countries where recent trends indicate a reduced allocation of national development budgets to agriculture, particularly in extension services and training, research, marketing and infrastructure. Furthermore, there has also been a decline in multilateral lending and bilateral aid for the agricultural sector.

It is a paradox that the strongest political lobbies in Washington, Brussels, and Tokyo happen to be the farming communities, whereas in many developing countries, the rural agricultural communities have little political voice in spite of the fact that agriculture often accounts for a very significant share of the gross domestic product.

The key question is how to link agricultural sciences, land, water, biodiversity management, technology and management, national and international policies and implementation actions across all stages of the food and agricultural chain, from production to marketing and consumption. Such an analysis cannot be carried out without reliable, timely and adequate data covering all aspects of the above linkages. The spatial and temporal rapidity of changes in the agricultural sector requires a wide range of timely and comprehensive data and information. The recent developments in remote sensing and Geographical Information Systems in combination with traditional ground-based information gathering and monitoring systems provide a basis for reliable, comprehensive and timely information relevant for decision-making.

The scientific community, civil society, national governments, and the international development community bear the fundamental responsibility for achieving a nutritionally healthy and productive and sustainable food agriculture, a goal which is also influenced by the increasing trend of the globalisation of the world's food and agriculture system.

4.2.2. Globalisation and food security

The challenge of food systems and globalisation[66,67] in the 21st century will require comprehensive social, economic, technological, and environmental evaluations and knowledge at the local, national, and international level to design policies that enable progress towards a food-secure world.[68]

The driving forces behind changes in food consumption include demographic changes, urbanisation, growing incomes, changing consumer preferences, international food marketing campaigns, and trade. The increasing global role and influences of transnational food corporations – as producers, processors, distributors and retailers – are resulting in significant changes in food consumption patterns, including the marketing of more and more processed foods, often with unhealthy levels of fat, salt and sugar contents. Consumers are also increasingly separated from the food production systems. They need the knowledge and education to make informed food choices. This includes ethical, moral, welfare, as well as economic, environmental, and food safety considerations.

4.2.2.1. Agricultural knowledge and food security

Policy and decision-making on the basis of a spatial and integrated agro-ecological and socio-economic policy analysis is essential for achieving sustainable world-wide agricultural development and food security. This requires a wide range of geographically specific and temporal information on agricultural land and water resources, weather and climate, agricultural practices as well as a vast array of socio-demographic data. Recent developments in remote sensing technology, Geographic Information Systems (GIS), computer and communications technology in combination with on-the-ground surveys and data gathering provide reliable, comprehensive and timely information for policy insight and implementation actions for local, national, regional and global food security and sustainable agricultural development.

The promise of science and technology cannot be delivered unless science is relevant and applicable to real, practical, and people-centred issues. The scientific and development policy community at the national level as well as the international level must work expeditiously towards the goal of achieving food security for all. This will require a systemic combination of the relevant sciences, including biological and biochemical, agro-ecological and environmental, social and economic science as well as informatics and knowledge communication.[69]

Food production systems interact with land and water resources and forest ecosystems, and biodiversity and environmental change will impact on all these inter-dependent systems. Ensuring soil fertility, genetic diversity, and agricultural water resource management is just as critical to increasing agricultural production as is adapting to the impacts of climate change. Agro-ecological sciences, together with the recent developments in geographical information systems including real-time remote sensing and the increasing quality and spatial coverage of sub-national, national and global resource data bases of soils, water resources, land cover, thermal regimes and rainfall patterns, population distribution etc., provide a sound scientific basis for productivity assessment as well as tools for identifying potentials and environmental constraints to crop production at the regional and national levels.

At the national level, governments need to invest in facilitating participatory and transparent utility-oriented knowledge systems that enable farmers to adopt sustainable food production systems. Agricultural knowledge systems should also give particular attention to integrating modern and traditional knowledge. At the international level, the agricultural knowledge system must facilitate the exchange and sharing of local, national and regional information and experiences. In this context, the international agricultural research community has a particular responsibility for the training and capacity building in developing countries. This is particularly important for new agricultural research and technology that is highly capital and knowledge-intensive.

The scientific challenges of sustainable development[70,71] and producing the food needed worldwide in the 21st century will require targeted and prioritised agricultural research with the participation of all stakeholders from producers to consumers. Agricultural research focusing on the differential vulnerability of farmers as well agricultural ecosystems will be required. Additionally, the threats of global environmental change need to be explicitly considered in mobilising agricultural research efforts, especially since a long time horizon is required from the initiation of research to local farm-level implementation. The science challenge ahead is a formidable one and can only be met through national and international commitments and investments harnessing the potential and utility of science, including:

- *Using science responsibly* – the emerging scientific tools of cellular and molecular biology can shorten the time and cut the costs required to develop innovative food varieties. Biotechnology tools can introduce genes that counter soil toxicity, resist insect pests, and increase nutrient content. Still, the questions of determining appropriate levels of risk and the ethics and societal acceptance of manipulating genetic material need to be resolved before the potential of biotechnology and genetic engineering can be realised.
- *Integrating ecology and socio-economy* – the progress in understanding the functioning of ecological systems, the compilation of an agricultural resource data base at the sub-national, national and global levels, and the development of analytical and mathematical modelling tools will be critical for a spatially relevant application of the results of agricultural research to for ensuring that the best choices are made at the sub-national level in the context of national needs within a world food economy.
- *Ensuring ecological sustainability* – new scientific tools will need to be combined with knowledge about natural resources in order to ensure sustainable and productive use and to avoid inefficient water use, the loss of arable lands, productivity declines, deforestation, pollution and the destruction of ecologically critical watersheds, the loss of biodiversity, and the health and environmental risks of intensive livestock production and fish farming.

- *Harnessing the Information Revolution* – the phenomenal potential of the information and communication revolution including remote sensing and GIS, the Internet, computer and communications technology, etc. can enable interactive agricultural research and sustainable development planning as well as real-time monitoring through ground-based and remote-sensing techniques.

4.2.2.2. New developments in agricultural remote sensing and GIS

Remote sensing and Geographic Information Systems technologies are relevant to the management of agricultural resources in terms of spatial, temporal, spectral and radiometric resolutions. For example, remote sensing techniques play an important role in crop identification, acreage and production estimation, disease and stress detection, and soil and water resources characterisation, by providing the required inputs with regard to developmental plans for land and water resources, by bringing additional land into cultivation through mapping and reclaiming wastelands, or by increasing the irrigation potential through groundwater prospects mapping. Other contributions of remote sensing and GIS include crop-yield and crop-weather models, integrated pest management, watershed management, agro-meteorological services, precision farming as well as social and economic policy analysis related to food security, poverty alleviation and sustainable agricultural development.

The application of remote sensing,[72] Geographic Information Systems and precision farming techniques in the management of agricultural resources are increasing rapidly due to improvements in space science supported by computer and communications technologies. The following challenges, for example, need to be addressed in the application of these technologies:

- Identification of crops and spatial estimation of crop area, yield and production.
- Forecasting of extreme events such as droughts and floods and using this information in a timely manner to establish early warning and response systems.
- Detection of crop stress due to nutrients, pests and diseases and quantification of their potential effects on the crop yield.
- Developing a precision farming database for smaller farm size and/or diverse crops/cropping systems.
- Developing decision support systems for the management of biotic and abiotic stresses at the farm level.
- Estimation of the depth of water in reservoirs and quality assessment of the ground water.

- Use of remote sensing and precision farming technologies in intercropping/ multiple cropping situations.
- Identifying ways and means of reducing time gaps in the collection, interpretation and dissemination of data to make them usable.

The traditional system of collecting and maintaining land use records is fraught with difficulties, and time-consuming ground surveys preclude timely monitoring and actions. Using remotely sensed data would enhance the accuracy and transparency in this process and particularly facilitate real-time monitoring. Remotely sensed data can guide farmers towards the right type of crops that can be grown given the existing soil conditions, thereby maximising output. Other information such as the availability of water for irrigation and the water-retaining capacities of the soil could also play a vital role in the choice of crops and inputs for maximum productivity.

By the remote surveying of agricultural areas for various factors like soil nutrient profile, water availability, crop status, etc., specific data could be generated for different agro-climatic zones and contribute to the planning of optimal cropping patterns. In addition, the assessment of production estimates based on the data provided would help farmers to minimise their losses.

4.2.2.3. Human well-being and the utility of remote sensing

In spite of numerous international commitments towards achieving food, water, and health security in the developing world over the last three decades, some 820 million people remain undernourished, over 1.2 billion lack access to safe water, and more than 2.5 billion are threatened by infectious diseases.

The world's population is projected to increase by some 50% to 9 billion over the next five decades, and this combined with economic growth and increasing consumption and pollution will result in substantial ecological pressures on land, water, atmosphere, and nature's biological resources. Global environmental change, including extreme weather events, climate change and climate variability will in the short and long term lead to agro-ecological changes that are likely to reduce productivity as well as the availability of arable land and water resources, particularly in areas where the majority of the most vulnerable populations live.

There is a critical need for spatial and temporal assessments and analysis of the differential vulnerabilities of ecosystems and populations[73,74] to target policy decisions regarding the achievement of food, water and health security, which are fundamental to achieving human well-being. Such assessments over wide geographical areas require not only real-time information, but also time series data to analyse and develop short and long-term integrated responses and development plans. Information from traditional on-the-ground surveys and observations is not only costly to collect, but delays in compiling and processing this information often inhibit timely response actions.

Remote sensing[75] has demonstrated the capacity and utility to generate essential and timely information with regard to changes in food production, emerging water scarcities and the monitoring of environmental changes that give rise to the emergence and spread of disease-causing pathogens and parasites. With regard to food security planning, remote sensing permits the identification and spatial assessment of geographically wide areas where a decline in food production is likely due to evolving environmental conditions, and also of areas where there may be food surpluses. Moreover, remote sensing makes other relevant information available such as the conditions and accessibility of the transport and storage infrastructure.

The systemic combination of remote sensing information with other on-the-ground socio-economic information on population, food stocks, etc. is important in the context of the timely analysis and planning of cost-effective responses, for example food aid deliveries and viable regional purchasing. In this context, the United States Agency for International Development's Famine Early Warning System[76] and the Food and Agriculture Organization of the United Nations' Global Information and Early Warning System on Food and Agriculture[77] are such operational systems providing invaluable service in emergency situations. These systems also have the potential to be further developed in combination with modelling and analytical tools for facilitating the longer-term planning of sustainable agricultural development and food security in response to, for example, future global climate and agro-ecological changes.

In the case of water resources planning, remote sensing has been shown to be relevant[78] to the repetitive monitoring of the variations in the surface elevations of lakes, rivers, wetlands and floodplain water resources. Such information in combination with other on-the-ground statistical and observational information provides for a sound basis for a sustainable national and regional development of water resources.

The remote monitoring of evolving ecological and environmental conditions such as the climate, soils and vegetation, water bodies and pollution, air quality, etc. combined with information on the distribution and movements of human populations and livestock provides a sound basis for the assessment and forecast of the potential emergence and spread of disease-causing pathogens and parasites.[79] Such knowledge is critical for health planners who often work with limited budgets and lack expert personnel to implement timely measures to contain the spread of human and livestock diseases. The growth in human and livestock populations together with global climate and ecological changes in the 21st century are also likely to exacerbate the risks of the emergence of new diseases.

The development and successful application of remote sensing requires ensuring that the information generated is relevant with regard to the identification and interpretation of the right mix of variables, geographical coverage and monitoring over time, etc. Also, the combination of remote sensing information with on-the-ground environmental, social and economic time series and observational information is important for facilitating modelling and policy analysis and generating knowledge in a form that can be communicated to policy makers for effective decision-making and timely implementation.

While remote sensing has substantial potential to enhance human well-being in terms of food, water and health security, the necessary infrastructure and expertise is lacking in many developing countries which are most threatened by global environmental change. There is an urgent need for an international partnership for establishing the infrastructure for remote sensing systems and training expert personnel.

4.2.3. Food security and climate change: an integrated agro-ecological and socio-economic world-wide assessment

The sustainability of land and water resources is significant to a range of themes and issues central to the study of global environmental change. Alterations in the Earth's surface hold major implications for the global radiation balance and energy fluxes, contribute to changes in biogeochemical cycles, alter hydrological cycles, and influence ecological balances and complexities. These environmental impacts at the local, regional and global levels, driven by human activity, have the potential to significantly affect food and water security and the sustainability of agro-ecological systems.

The combination of a spatially detailed biophysical/agronomic assessment tool and a global food system model provides an integrated ecological–economic framework for the assessment of the impacts of climate change and agricultural vulnerability. The focus has been on analysing the current and future availability and use of regional and global land resources in the context of technological and environmental changes, including climate change and climate variability.

The Food and Agriculture Organization of the United Nations (FAO) and the International Institute for Applied Systems Analysis (IIASA) has developed integrated ecological–economic analytical tools and global databases over the

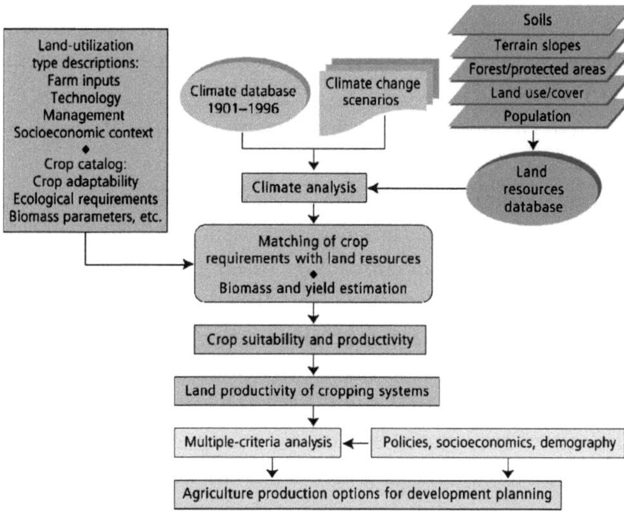

Fig. 1. *Agro-ecological zone methodology.*

last two decades. The focus has been on multi-disciplinary scientific research analysing the current and future availability of land and water resources in the face of local, national and supra-national demographic, socio-economic, international trade and globalisation and technological and environmental changes, including climate change and climate variability.

The Agro-Ecological Zone (AEZ) modelling framework[80] shown in Figure 1 synthesises essential components of both the crop and ecosystem models described above. It uses detailed agronomic-based knowledge to simulate the availability and use of land resources, farm-level management options, and crop production potentials; at the same time, it employs detailed spatial biophysical and socio-economic ground-based and remote sensing datasets to distribute its computations at fine-gridded intervals over the entire globe. This land-resource inventory is used to assess, for specified management conditions and levels of inputs, the suitability of crops in relation to both rain-fed and irrigated conditions, and to quantify the expected attainable production of cropping activities relevant to specific agro-ecological contexts characterising the study area. The characterisation of land resources includes components of climate, soils, landform, and present land cover.

Crop modelling and environmental matching procedures are used to identify crop-specific environmental limitations under various levels of inputs and management conditions. Specifically, AEZ employs the FAO/UNESCO Digital Soil Map of the World (DSMW) as the underlying reference for its own land surface database, consisting of more than 2.2 million grid cells at $5' \times 5'$ latitude/

longitude, i.e. with a size of about 10 km×10 km at the equator. In addition, a global digital elevation map (DEM) and a derived slope distribution database is linked to DSMW.

AEZ's current climate database is based on the Climate Research Unit (CRU) of the University of East Anglia which consists of historical monthly mean data (mean monthly minimum temperature, mean monthly maximum temperature, precipitation, solar radiation, vapour pressure deficit) for the period 1901–1996, and includes a monthly-mean climatology based on the decades 1960–1990. In AEZ, the CRU data are transformed into daily data and analysed vis-à-vis crop requirements. Finally, AEZ employs a land cover/land use layer specifying distributions of aggregate land-cover classes, as derived from global 1-km land-cover datasets from NOAA AVHRR and GLC2000, respectively.

The AEZ global land resources database incorporates spatial delineation and the accounting of forest and protected areas. In terms of key socio-economic datasets, AEZ employs a global population data set calibrated for the year 2000, including estimates of the spatially explicit population distribution and density for each country.

In summary, the AEZ framework contains the following basic elements:

- Land resource database, containing geo-referenced climate, soil and terrain data;
- Land Utilization Types (LUT) database of agricultural production systems, describing crop-specific environmental requirements and adaptability characteristics, including input level and management.
- Mathematical procedures for matching crop LUT requirements with agro-ecological zones data, including potentially attainable crop yields estimates, by land unit and grid-cell (AEZ global assessment assesses 2.2 million grid cells, covering a 5′×5′ latitude/longitude grid, based on a 1:5,000,000 scale global soil map);
- Assessments of crop suitability and land productivity; and
- Applications for agricultural development planning.

The world agricultural economy and trade model (Basic Linked System – BLS)[81] developed at the International Institute for Applied Systems Analysis (IIASA)[82] comprises a series of national and regional agricultural economic models.

It provides a framework for analysing the world food system, viewing national agricultural components as embedded in national economies which in turn interact with each other at the international trade level. The BLS model consists of 34 national and regional geographical components covering the globe, calibrated and validated over past time windows, and successfully reproduces the regional consumption, production and trade of major agricultural commodities in 2000.

The individual national/regional models are linked together by means of a world market where international clearing prices are computed to equalise global demand with supply. Each individual model component focuses primarily on the agricultural sector, but attempts to represent the whole economy as necessary for capturing essential dynamics among capital, labour and land. For the purpose of subsequent international linkage, the production, consumption and trade of goods and services are aggregated into nine main agricultural sectors, though individual regional models have more detail. The nine agricultural sectors include: wheat; rice; coarse grains; bovine and ovine meat; dairy products; other meat and fish; protein feeds; other food; and non-food agriculture. All physical and financial accounts are balanced and mutually consistent: the production, consumption, and financial accounts at the national level, the trade and financial flows at the global level.

Within each regional unit, the supply modules allocate land, labour and capital as a function of the relative profitability of its different economic sectors. In particular, the actual cultivated acreage is computed from agro-climatic land parameters (derived from AEZ) and profitability estimates. Once acreage, labour and capital are assigned to cropping and livestock activities, actual yields and livestock production are computed as a function of fertiliser applications, feed rates, and available technology.

Population growth and technology are key external inputs in the BLS. Population numbers and projected incomes are used to determine the future demand for food. Technology affects BLS yield estimates by modifying the efficiency of production per given units of input. For simulations into the future, scenarios of socio-economic development and population growth must be chosen in order to inform BLS computations. Another key external input to AEZ/BLS is the climate and environment, which determines crop suitability and the potential yields used by the economic model as an input in resource allocation. Thus, projected climate change affects BLS results indirectly yet significantly via its impacts on agro-climatic land resources computed by AEZ.

In order to assess agricultural development over this century, with or without climate change, it is necessary to first make some coherent assumptions about how key socio-economic drivers of food systems might evolve over the same period. To this end, we used plausible socio-economic development paths, as specified in the IPCC Special Report on Emissions Scenarios (SRES). The SRES scenarios have been constructed to explore socio-economic development and related pressures on the global environment in this century, with special reference to the emission of greenhouse gases into the atmosphere.

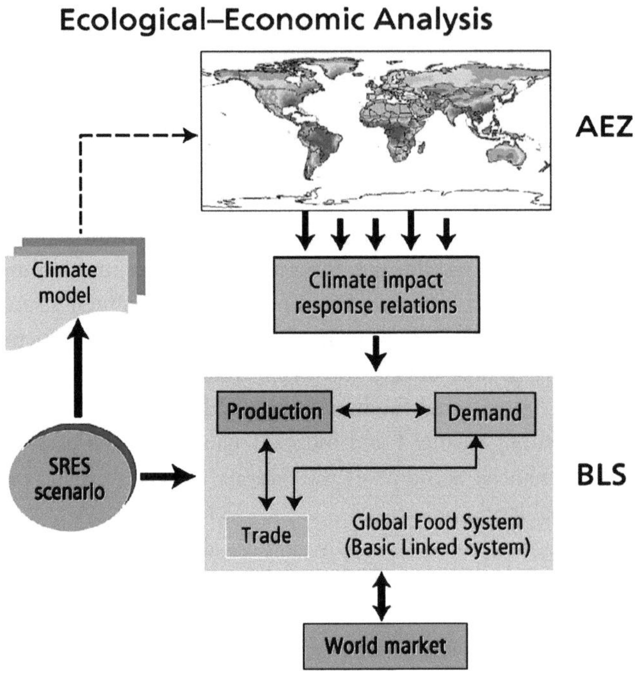

Fig. 2. *Integrated ecological–economic analysis (source: Fischer et al., 2002b).*

Within this context, climate change is clearly seen as the consequence of complex social, economic, and environmental interactions, possibly modulated by the regional and global capacity to mitigate and adapt. Emissions of greenhouse gases connected to specific SRES scenarios are translated into projections of climate change over this century by using general circulation models (GCMs).

The integrated methodology and database developed, as shown in Figure 2, provides a foundation for detailed country studies. The main results of the IIASA study include climate-change impacts on the prevalence of environmental constraints on crop agriculture; climate variability and the variability of rain-fed cereal production; changes in potential agricultural land; changes in crop production patterns; and the impact of climate change on cereal production potential. Results of the AEZ–BLS integrated ecological–economic analysis of climate change on the world food system include the quantification of the scale and location of hunger, international agricultural trade, prices, production, land use, etc. The analysis assesses trends in food production, trade, and consumption, and the impact on poverty and hunger of alternative development pathways and varying levels of climate change. Some highlights of the results are summarised below.

4.2.3.1. Summary of results

The AEZ–BLS results of climate change impacts on spatial agro-climatic resources, land and crop productivity, cereal production, risks of hunger, agricultural GDP and international trade, agricultural research, mitigation and adaptation priorities are summarised below.

Climate change constraints on crop cultivation: Two-thirds of the global land surface – some 8.9 billion ha – suffer rather severe constraints on rain-fed crop cultivation: 13.2% are too cold, 26.5% are too dry, 4.6% are too steep, 2.0% are too wet, and 19.8% have poor soils. Climate change will have both positive and negative impacts, as some of these constraints will be alleviated while others may increase. In terms of negative impacts, arable land will suffer severe environmental constraints, preventing crop production in Central America and the Caribbean (by up to 3%), in Oceania and Polynesia (by up to 5%), in Northern Africa (by up to 3.5%), Eastern Africa (by up to 2.4%), Central Africa (by up to 1.2%) and Western Africa (by up to 1.7%) in the 2080s. Southern Africa will be the most severely affected region with approximately 11% of its total land area (265 million hectares) at risk of being lost for crop production due to the environmental constraints induced by climate change.

The world's boreal and arctic ecosystems are likely to decline by 60% due to a northward shift of the thermal regimes. The semi-arid and arid land areas in developing countries may increase by about 5–8%. The agro-ecological changes due to climate change will also result in water deficits in some areas and surpluses in others as well as in the increased or reduced infestation of disease pathogens and parasites.[83]

Climate change impact on cereal production: a clear distinction emerges between the "gainers" and "losers" of the impact which climate change will have on agricultural production. Climate projections predict a considerable increase in land suitable for cereal production in the developed nations. Increases will predominantly occur in North America (a 40% area increase of the currently 360 million hectares of cultivated land), in Northern Europe (a 16% area increase of the 45 million hectares currently being cultivated), in the Russian Federation (a 64% area increase of the currently 245 million hectares), and in East Asia (a 10% area increase of the 150 million hectares presently under cultivation).

Currently, there are some 78 food-insecure countries with a current population of 4.2 billion, of which 18% are undernourished. The total population of these countries in the 2080s is projected at 6.8 billion. Of particular concern are some 40 developing countries – many of them in the least developed country group – that may lose some 10–20% of their cereal production capacity due to climate change by the 2080s.

According to the HadCM3 model's climate change projections, 27 countries will lose 2.2% and 52 countries will lose 5% of their cereal production potential by 2080 due to climate change. In turn, the cereal production potential of 42 and 59 countries will increase by 4% and 6%, respectively.

Climate change and number of people at risk of hunger: some fairly robust conclusions emerge from the analysis of climate-change impacts on the number of people at risk of hunger. First, climate change will most likely increase the number of people at risk of hunger. Second, the importance and significance of the climate-change impact on the level of undernourishment depends entirely on the level of economic development assumed in the SRES scenarios. Figure 3 summarises the simulation results, showing the additional number of people at risk of hunger in 2080 plotted against different levels of atmospheric CO_2 concentrations and associated climate changes, and Figure 4 shows the additional number of undernourished by selected regions and climate modes and IPCC future development scenarios.

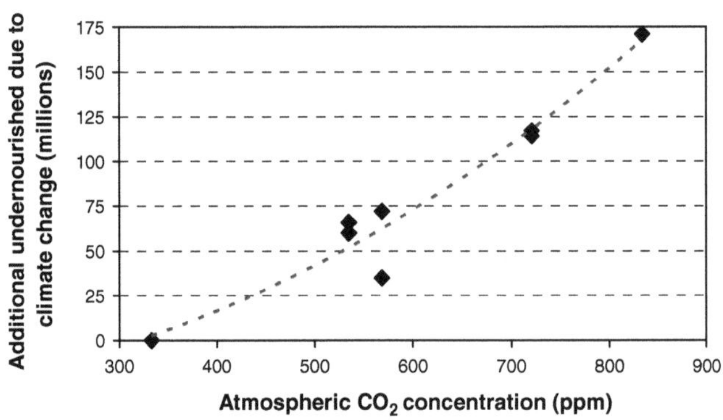

Fig. 3. *Increase in undernourished persons due to climate change (source: Fischer et al., 2002b).*

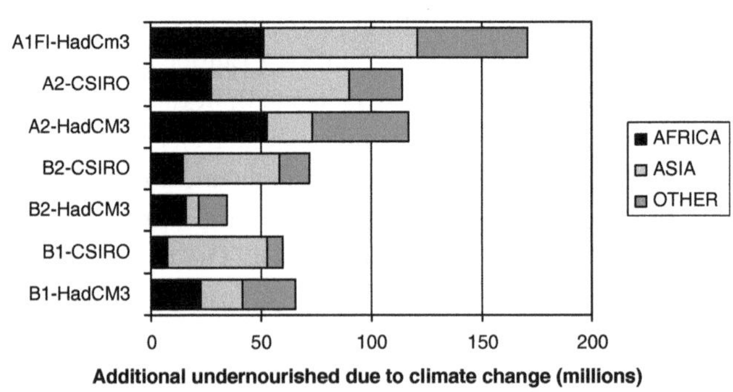

Fig. 4. *Additional undernourished persons due to climate change (source: Fischer et al., 2002b).*

The number of undernourished in the developing world is estimated at around 820 million, equivalent to 17% of the total population of 4.1 billion. Sub-Saharan Africa has the highest incidence of undernourishment, with roughly 32% of the total population being deprived of access to food. In 2080, climate change will have resulted in an additional 35–170 million undernourished people in developing countries; SSA will be the region most affected, with an additional undernourished population of 17–50 million.

Climate change impact on agricultural GDP: the impact of climate change is relatively small at the aggregate global level (between −1.5% and +2.6%). This applies to the total global GDP of agriculture in the reference projections, ranging from 2.9 to 3.6 trillion U.S. dollars (at 1990 prices). Large variations are expected to occur between regions. Developing regions, with the exception of Latin America, will confront negative impacts on agricultural GDP. By 2080, climate change will reduce Asia's agricultural GDP by 4% and Sub-Saharan Africa's by up to 8%. According to scenario A2, North America will gain between 3% and 13% of agricultural value added; Western Europe will lose between 6% and 18%, and the former Soviet Union will gain between 0% and 23%.

Climate change impact on international cereal trade: baseline scenarios indicate a growing dependence of developing countries on net cereal imports, ranging between 170 and 430 million tonnes. In several developing countries, between 10% and 40% of the cereal consumption will have to be covered by imports. Many of these countries, however, lack the foreign exchange to finance food imports, which puts them at risk of increased food insecurity. At present, the SSA region's net cereal imports amount to approximately 7 million tonnes, but the impact of climate change may result in a net import of roughly 143 million tonnes of cereal by 2080.

Climate change impacts and issues of fairness and justice: global environmental change raises the issue of fairness and equity. Climate change is global and long-term and involves complex interactions between climatic, environmental, economic, political, institutional, social and technological processes. It has significant international and intergenerational implications in the context of equity and sustainable development. Developing countries have thus far contributed relatively little to the causes of climate change. Sub-Saharan Africa's cumulative CO_2 emissions over the last 50 years total about 2% of the world's aggregate CO_2 emissions. With 9.7% of the world's population, SSA at present only contributes 2.4% of the world's total CO_2 emissions. And yet, this region will bear the brunt of the impacts of climate change.

Climate change mitigation: the cumulative green house gas emissions of the past 50 years already in the atmosphere will, in all certainty, lead to climate change in the next five decades. This will substantially impact the viability and productivity of agricultural areas. Developed countries, accounting for about one-fifth of world's current population, have contributed over 70% of the CO_2 emissions accumulated in the atmosphere over the last 50 years. Developing countries have thus far contributed relatively little to the causes of climate change. Yet many of the poorest countries with the largest food deficit will suffer substantial losses in domestic food production, which will contribute to increased food insecurity and hunger. This situation raises a fundamental concern of fairness and justice, as many developing countries have contributed very little to the causes of climate change and yet will bear the brunt of the negative impacts of climate change in the form of wide-spread economic losses and further-exacerbated chronic hunger and debilitating poverty.

The United States' current per capita emission of CO_2 amounts to some 5.5 metric tonnes compared to Europe's 2 metric tonnes and South Asia's and Latin America's 0.5 and 0.7 metric tonnes, respectively. The developed countries need to urgently adopt measures to reduce emissions, while the developing countries need to rethink their energy development polices and adopt technologies that mitigate climate change. At the same time, the highest priority needs to be given to the adaptation to climate change, especially in the poor developing countries which often have a low capacity to do so.

Adaptation to climate change: national governments in developing countries, together with their bilateral and multilateral development partners, must devise mechanisms that provide climate information and forecasting, scientific research and policy analysis at the national and regional levels to adapt to climate change. Adaptation to climate change requires effective partnerships between the public and private sectors with the operational participation of all stakeholders, including governments, farmers, the scientific and research community, NGOs, and private business. Due to the long time-lag between the development of adaptation strategies and technologies and their actual implementation, investments and sustained international funding will be essential to the promotion and strengthening of comprehensive climate information systems and applied climate research for policy actions, for the mobilisation of funding for climate adaptation, and the establishment and prioritisation of agricultural research and agricultural knowledge systems as well as agricultural extension and marketing services.

4.2.4. Concluding remarks

The climate change issue is global, long-term and involves complex interaction between demographic, climatic, environmental, economic, health, political, institutional, social and technological processes. It has significant international and intergenerational implications in the context of equity and sustainable development. Climate change poses a substantial risk to the sustainability of agriculture and the most critical need for making progress towards achieving world-wide food security relates to the availability of reliable, comprehensive and timely information at the local, national, regional and global levels. The recent developments in remote sensing, computer and communications technology in combination with traditional data gathering practices provide a sound scientific basis for real-time monitoring, policy analysis and decision-making towards achieving food security and sustainable agricultural development.

---

[63] Food and Agriculture Organization. The State of Food Insecurity in the World. Rome: FAO, 2006.

[64] Food and Agriculture Organization; World Health Organization. Nutrition and Development: A Global Assessment. International Conference on Nutrition, Rome, Italy, 1992.

[65] United Nations Millennium Project. Investing in Development: A Practical Plan to Achieve the Millennium Development Goals. New York: UN Millennium Project, 2005.

[66] Shah, Mahendra M. "Globalization and Food Systems: A Holistic Perspective." IHDP Update 1 (2005): 6–8.

[67] United Nations Population Fund. Global Population and Water: Access and Sustainability. Population and Development Strategies 6. New York: UNFPA, 2003.

[68] Shah, Mahendra M. "Food in the 21st Century: Global Climate of Disparities." GAIA – Ecological Perspectives for Science and Society 10.4 (2001): 273–278.

[69] Shah, Mahendra M. and Strong, Maurice. Food in the 21st Century: From Science to Sustainable Agriculture. Washington, D.C.: World Bank, 2000.

[70] United Nations Conference on Environment and Development. The Global Partnership for Environment and Development: A Guide to Agenda 21. Geneva: UNCED Secretariat, 1992 (co-ordinating author: Mahendra M. Shah).

[71] Shah, Mahendra M. "Humanity, Nature, and Sustainable Development." The Wealth of Diversity: The Role of Universities in Promoting Dialogue and Development. Highlights of the 12th IAU 2004 General Conference, São Paolo, Brazil. Paris: International Association of Universities; São Paolo: University of São Paolo Press, 2005.

[72] Patil, W.C., Maru, Ajit, Shashidhara, G.B., and Shanwad, U.K. "Remote Sensing, Geographical Information System and Precision Farming in India: Opportunities and Challenges." Proceedings of the Third Asian Conference for Information Technology in Agriculture, Beijing, China, 26–28 October 2002. 478–483.

[73] Shah, Mahendra M., Fischer, Günther and van Velthuizen, Harrij. "Food, Water, Health, and Infectious Diseases: Focus on Global Change." Interactions between Global Change and Human Health. ed. Pontifica Academia Scientiarum. Vatican City: Pontifica Academia Scientiarum, 2006. 230–251.

[74] Lutz, Wolfgang and Shah, Mahendra M. "Population Should Be On the Johannesburg Agenda." Nature 418.6893 (2002): 17.

[75] National Research Council (U.S.). Contributions of Land Remote Sensing for Decisions about Food Security and Human Health: Workshop Report. Washington, D.C.: National Academic Press, 2007.

[76] See: Famine Early Warning Systems Network. www.fews.net

[77] Food and Agriculture Organization. "The Global Information and Early Warning System on Food and Agriculture." 1 Dec. 2007. http://www.fao.org/giews/english/giews_en.pdf

[78] Shakoor, A., Shehzad, A., and Asghar, M.N. "Application of Remote Sensing Techniques for Water Resources Planning and Management." Paper presented at the 1st International Conference on Advances in Space Technologies, Islamabad, Pakistan, 2–3 Sept. 2006.
[79] Beck, Louisa R., Lobitz, Bradley M., and Wood, Byron L. "Remote Sensing and Human Health: New Sensors and New Opportunities." Emerging Infectious Diseases 6.3 (2000): 217–226.
[80] Fischer, Günther, van Velthuizen, Harrij, Shah, Mahendra M., and Nachtergaele, Freddy O. Global Agro-Ecological Assessment for Agriculture in the 21st Century. Laxenburg: International Institute for Applied Systems Analysis, 2002.
[81] Fischer, Günther, Shah, Mahendra M., and van Velthuizen, Harrij. Climate Change and Agricultural Vulnerability. Laxenburg: International Institute for Applied Systems Analysis, 2002.
[82] IIASA is an international research organisation. It conducts inter-disciplinary scientific studies on environmental, economic, technological, and social issues in the context of the human dimensions of global change. IIASA is located in Austria near Vienna and is sponsored by its national member organisations in Africa, Asia, Europe, and North America.
[83] Fischer, Günther, Shah, Mahendra M. and Rollinson, David H.L. Agricultural Perspectives in the Tsetse Infested Areas in Africa. IIASA/FAO Collaborative Paper CP-85-12. Laxenburg: International Institute for Applied Systems Analysis, 1985.

4.3 The use of space technologies for effective and sustainable responses to food insecurity threats in Africa

Menghestab Haile

4.3.1. Overview: food insecurity

In Africa, despite its vast reserves of natural resources and agricultural potential, millions of its population are currently chronically exposed to food insecurity and to recurrent food emergencies. Declining agricultural productivity, deepening rural poverty, inadequate investment in rural infrastructure, ineffective marketing strategies, a lack of employment opportunities, conflicts, and increased disease burdens such as HIV/AIDS compounded with extreme climatic events are reducing rural populations' ability to access food and sustain their livelihoods.

Recurrent extreme weather events and climate change have a direct impact on agricultural productivity and rural incomes, resulting in chronic hunger and food shortages. Faced with recurrent food shortages, rural communities are expanding their areas under cultivation, often into non-arable marginal areas that are unsuitable for cropping. This accelerates land degradation, soil erosion and soil infertility, further increasing these communities' vulnerability to agricultural failure. Environmental degradation is further accelerated due to the increased deforestation to create more agricultural land, wood fuel and other wood products. Competition over resources such as pasture and water as a result of expanding cultivation into traditionally pastoral areas is increasing the potential for conflicts between pastoralists and sedentary agriculturalists.

World leaders, national governments, the humanitarian community and donor agencies have made commitments at various international summits and conferences to address the root causes of poverty and chronic hunger and reduce human suffering. The Millennium Declaration of the United Nations Millennium Summit in September 2000, where world leaders adopted the Millennium Development Goals (MDGs), has set clear targets for reducing poverty, hunger, diseases, illiteracy, environmental degradation, and discrimination against women by 2015. Subsequently, the Johannesburg Declaration on Sustainable Development of September 2002 affirmed the objectives set by the World Food Summit and recognised that sustainable agriculture and rural development are essential for increasing food production and enhancing food security and food safety in an environmentally sustainable way. More recently, the G8 countries have taken concrete steps to meet their commitments regarding the reduction of poverty and elimination of hunger by agreeing to the cancelling of 100% of the debts of many African countries, mostly in Sub-Saharan Africa. Recently, the African Union and the EU held a joint summit and reaffirmed their commitment to addressing poverty in Africa.

The achievement of the MDGs would depend on the ability to plan and implement various poverty reduction programmes. This would depend on the availability of resources as well as the availability of reliable information on who the poor communities and households are; why they are poor and vulnerable; and what the main risks affecting their livelihood security are, so that appropriate intervention options can be designed and implemented. Reliable data is also required to monitor the results, showing if programmes are achieving the intended objectives. Poor countries lack the institutional capacity to regularly collect and analyse relevant data. In most cases, this is compounded by the lack of clear government policies and strategies to address poverty and risk management.

There is a need to build institutional capacity that would allow the effective monitoring of food security and vulnerability. This would involve improving the existing conventional data collection mechanisms in the relevant government departments and incorporating the recent developments in geospatial technologies.

As a result of the recent advances in satellite technologies, ICT and GIS applications, geospatial data is becoming more widely and readily available for civilian use. Satellite data, computing capacity, and GIS software are becoming more and more affordable and accessible. The humanitarian community is quickly adopting these latest technological developments and tools to improve the effectiveness of humanitarian interventions. Geospatial applications in the context of this paper encompass spatial data collection, geospatial analytical tools, and the storage and dissemination of geo-referenced information. In addition, advances in satellite communications have revolutionised information dissemination. The purpose of this paper is to document the current uses of space technology for supporting humanitarian operations in a more sustainable and affordable way.

4.3.2. The poverty reduction potential of space technologies

Space-borne systems and related technologies and applications are quickly developing, as new generations of satellites are rapidly becoming available for peaceful uses. Earth observation applications can provide advanced tools that enhance the collection, storage, analysis and integration of spatial/geographic data with related non-geographic information collected from ground surveys and stored in databases. Governments, development partners and the humanitarian community are making use of the various types of Earth observation products to support their development planning and field-level activities. Earth observation for humanitarian purposes is mainly used for food security and vulnerability monitoring, disaster assessments and operational planning. In summary, the following are the components of the food security knowledge base and understanding that would benefit from space technology:

- Baseline risk mapping – livelihood zoning, vulnerability profiling, agro-ecological zones, cropping patterns
- Agricultural monitoring – area cultivated, vegetation status, rainfall and yield indicators
- Disaster assessment – flooded area, drought-affected area, etc.
- Operational planning – infrastructure, accessibility, insecurity situation, water points.

The list of cartographic products provided below represents a first attempt to identify potential EO-derived products that could enhance the efficiency and effectiveness of poverty reduction and food security activities.

(1) Typical country-specific baseline information
   - Infrastructure maps (including roads, storage facilities, markets, physical accessibility)
   - Population maps (disaggregated by livelihood categories)
   - Topo/land form maps (including elevation)
   - Land cover, land use and land use change maps
   - Hazard and risk maps (e.g. landslides, flooding, flash floods, droughts)
   - Pastoral maps (including carrying capacity maps)
   - Forest maps (including fuelwood availability)
   - Soil maps (salt-affected soils).

(2) Typical country-specific information on environmental monitoring (agricultural production)
   - Acreage maps per administrative unit/agro-ecological zone/farming system
   - Crop type (food and cash crops) and crop distribution maps per administrative unit/agro-ecological zone/farming system

- Yield estimates (outlook, forecast) – multi-temporal products per administrative unit/agro-ecological zone/farming system
- Meteorological data (rainfall, radiation, relative ETP) – multi-temporal products
- Vegetation vigour (NDVI or other indices) – multi-temporal products
- Start and end of the cropping season per administrative unit/agro-ecological zone/farming system
- Cropping season development status index maps per administrative unit/agro-ecological zone/farming system
- Crop yield outlook (three times per season) maps per administrative unit/agro-ecological zone/farming system
- Biomass availability (e.g. forage availability maps) per administrative unit
- Snow cover change maps – periodic snow melting maps (mountainous regions, e.g. Afghanistan, Nepal, Mongolia, etc.) – water availability forecasts
- Extent of bush-forest burned maps – whenever needed.

(3) Typical country-specific information to improve disaster preparedness (disaster risk forecast maps)
- Synoptic weather forecast (early warning)
- Disaster risk maps
- Disaster risk impact maps
- Land slides: stability of slopes (land displacement)
- Population movement
- Physical accessibility
- Physical vulnerability maps (chronic vulnerability) – composite maps.

(4) Typical country-specific information to improve relief operations
- Topo/land forms maps
- Population distribution maps
- Climate and climate forecast maps (including rainfall/snow distribution)
- Maps showing the existing infrastructure and disaster-related damages (e.g. existing roads, air fields, fuel distribution points, bridges, buildings, schools, hospitals, market places, etc.)
- Potential rescue-related maps showing potential food distribution points, warehouses and storage facilities, shelters and safe places, safe drinking water points
- Land use maps (including rivers)
- Extent of natural disaster damages maps (volcanic eruption, earthquakes, land slides, flash floods, major floods, hurricanes)
- Damage maps (resulting from natural disasters and violent conflicts/wars)
- Mine fields/de-mined areas maps.

Satellite technologies have greatly advanced communications. GPS and mobile phones make communication possible. Satellites are also used for data uplink and downlink facilities for data sharing. As a result of the improvement of communications, the exchange of information has become possible. Some of the areas that benefit from this technology include:

- Market information (improved connectivity where previously rural infrastructure did not allow for telephone lines). Farmers are using mobile phones to monitor the markets
- Improved education delivery
- Health delivery
- Disaster response
- Public information.

4.3.3. Space technologies for food insecurity risk analysis

A key component in establishing an effective food security risk management capacity is the ability to develop a comprehensive understanding of the various risk factors that impact on food security. Risks need to be identified, assessed and monitored. A comprehensive food security risk analysis and monitoring system should provide information that allows for answering questions such as: who are the food-insecure and vulnerable; where do they live; why are they food-insecure and vulnerable; what intervention options are most appropriate? In summary, it is important to identify and understand the various food security hazards and vulnerabilities at the sub-national level, identify knowledge gaps and national capacities, identify the temporal and spatial distribution of both historical, current and expected hazards, identify the magnitude, frequency and duration of the hazards, and assess which of the hazards can be reliably predicted.

In order to integrate the various aspects of hazards and vulnerabilities, appropriate risk assessment tools need to be developed. The tools will make it possible to integrate the possible impacts of various natural hazards such as floods, droughts and earthquakes with socio-economic and vulnerability factors. Risk assessment and monitoring comprises the following main elements: risk knowledge, hazard monitoring and early warning, needs assessment, and communication. The assessments of risks requires the systematic collection and analysis of various data sets and should take into account the dynamics and variability of hazards and vulnerabilities from processes such as urbanisation, rural land-use change, environmental degradation, and climate variability/change. In summary, the main steps in developing a comprehensive risk analysis include:

- Undertake risk identification and the development of risk, hazard and vulnerability maps through comprehensive food insecurity risk and vulnerability analysis;
- Establish food security risk monitoring systems based on the key indicators identified in (1) to provide early warning;
- Conduct needs assessments during emergencies;
- Develop a food security information management platform that incorporates the latest advances in GIS and spatial technology; and
- Support the capacity building of national and regional institutions.

Essentially, a food security risk management information system should provide the following:

- Identify geographical areas and communities that may be facing or will face immediate, acute food insecurity;
- Identify the main causes of food insecurity and risks to livelihoods and indicate the extent (or magnitude) to which households will be affected;
- Identify various risk management efforts that governments, communities and households should undertake to avoid or mitigate food insecurity situations;
- Provide opportunities for the early detection of risks and for addressing potential adverse impacts; and
- Trigger contingency planning, support advocacy and resource mobilisation efforts, and support market interventions such as the local purchasing of food.

4.3.3.1. Baseline mapping (identifying food security risks)

Food insecurity risk and vulnerability information can be presented in a livelihood atlas. Such an atlas can describe the vulnerability and hazard components of risk factors in a formal way through data, maps and analyses. The data would include environmental (climate) and political hazards and socio-economic vulnerabilities. The analysis includes an understanding of the coping capacities of people dealing with risks, thereby providing a better context for interpreting early warning information and the impact of shocks or hazards on transitory food insecurity. Livelihood zones and risk zones can then be established, along with composite vulnerability indices for the geographic targeting of zones at risk. Together with hazard and vulnerability maps, this would provide the basis for disaster preparedness strategies and the development of adaptation programmes. A livelihood atlas can provide the platform upon which monitoring systems can be developed. Monitoring systems would be based on indicators – satellite data, social factors (such as market analysis) and environmental factors (such as drought).

When undertaking a comprehensive food security baseline analysis, there are core analytical steps that are used. Below follows a summary of some of the key steps:

- Secondary data analysis: available data sets are used to undertake a preliminary vulnerability and hazard analysis. This requires the acquisition of all relevant secondary data including socio-economic and environmental data. A secondary data analysis will allow the development of household and community vulnerability profiles. It will identify geographical areas at risk as well as areas that may require further investigation through primary data collection. A key output of this process is the creation of geo-database at the national level and also the identification of priority geographical areas that need further study.
- Vulnerability, hazard and livelihood zoning: a secondary data analysis will allow for the production of vulnerability, hazard and livelihood zone maps at the subnational and national level. Hotspot areas that would require monitoring will be identified and the relevant indicators selected. The outputs depend on the availability of data.
- Primary data collection: very often, a secondary data analysis may not be adequate for answering all the relevant questions on food security. A primary data collection may be required to fill the gaps. A geospatial data analysis can then be used to refine the design of the survey. The use of PDA and GPS can improve the field data capture and analysis. By geo-referencing the data, they will be easily integrated when undertaking a vulnerability and risk analysis.
- Risk mapping: by integrating the secondary data analysis with the primary data analysis, hazard, vulnerability and risk profile maps are produced.
- Geospatial data management and dissemination: a geospatial information management and data exchange platform should be developed and maintained. All the relevant data and analyses will be managed and effectively shared with users.

In undertaking a risk analysis, the first step in this process would be to assess if there is any relevant work done at the country level and identify both national and international partners. It is also important to assess the local capacity and institutions in the area of remote sensing, GIS and risk analysis. A risk analysis framework

Tab. 1: *Monitoring (environment and agriculture).*

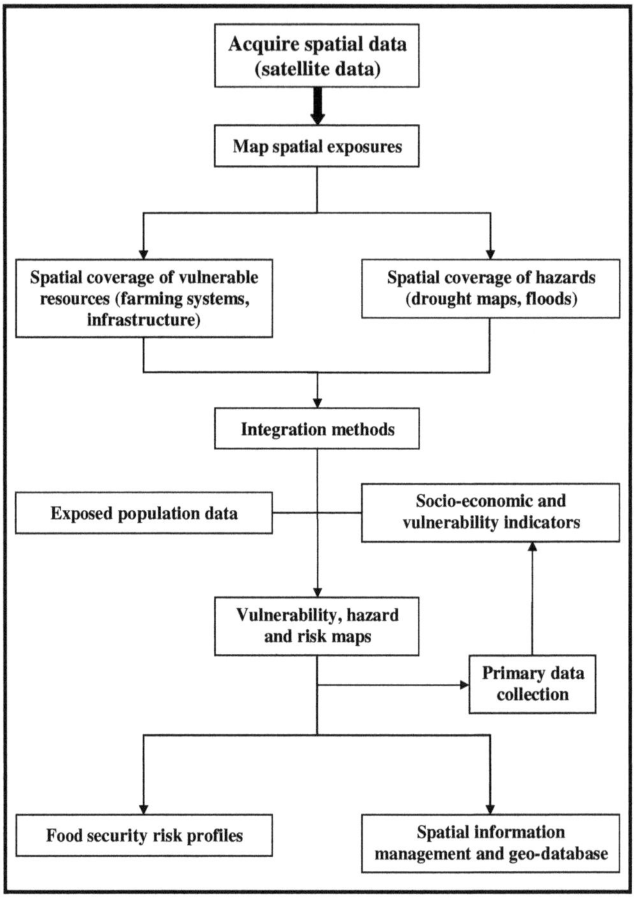

needs to be established, as well as a step by step process that includes the elements of spatial data identification, analysis and the dissemination of the results.

During the last few years, our climate monitoring and weather observation capacity has greatly improved due to the latest developments in the field of meteorological modelling, an increased satellite data availability, and advancements in ICT. Similarly, advances in crop modelling, improved cropped area estimation and an enhanced vegetation monitoring capacity have contributed to improvements in agricultural monitoring. The ability to better monitor weather and agriculture means that major agricultural and pasture failures can be detected early and agricultural losses can be predicted several months before the end of the growing season. In this way, humanitarian needs and resource requirements can be determined much earlier than was hitherto possible.

4.3.3.1.1. Agricultural monitoring

The agricultural sector is the main determinant of the food security situation in rural Africa. The key component of a food security information system is the agricultural monitoring framework which includes: risk and vulnerability analysis, seasonal forecasting, agro-meteorological monitoring, and post-harvest assessments.

4.3.3.1.2. Estimating the cultivated area

There is a need to improve the area cultivated in small-holding farming systems. USGS has been working on hybrid Landsat, MODIS, etc. methods for this with some initial positive results, but with a great distance yet to cover before getting an early-season, mid-season, and end-of-season picture of this factor.

4.3.3.1.3. Rainfall monitoring

For communities that are dependent on traditional rain-fed agriculture, rainfall is a key determinant of yield and production. Rainfall is also important for pastures. However, areas and communities that are affected by droughts and require humanitarian assistance often have the poorest rainfall station coverage. As a result, it has been difficult to monitor rainfall where we need it most. Satellite rainfall estimation techniques have been introduced to improve this situation.

4.3.3.1.4. NDVI monitoring

Vegetation monitoring using satellite data (NDVI) has been going on for more than 20 years now. Currently, the main sources of vegetation monitoring are NDVI (AVHRR) and SPOT Vegetation.

- SPOT data are distributed by VITO and at times, there is a processing and/or delivery delay. GMFS, FAO and WFP will see what can be changed to regularise the SPOT availability.
- Some research is going on for creating an inter-calibrated time series of AVHRR and SPOT data. The collaborating agencies are VITO (GMFS) and the South African Satellite Application Centre. There is a need to link this research with NASA. FAO and WFP will facilitate this link.
- There is a need to look at the new products such as MODIS and MERIS and how they can be used in conjunction with NDVI and SPOT. NASA/USGS and GMFS will follow up.

4.3.3.1.5. The Water Requirement Satisfaction Index (WRSI, crop condition)

This is a USGS-led application to monitor the daily condition of crops by looking at how well a specific crop's water requirements have been met in a specific area. This spatial application is derived from earlier WRSI work by FAO. This data will be made available to all parties involved in this effort. USGS will identify how to proceed.

4.3.3.2. Disaster assessment and operational planning

The latest space technological advances are playing a major role in undertaking effective disaster assessments when major disaster such as the tsunami of December 2004 hit. Satellites are being used both to carry out the actual disaster area assessment as well as to help with the operational planning that includes logistics assessments and communications support. Various space agencies have recognised the importance of working together by pooling the various space capabilities of their institutions during disaster times. As a result, the International Charter was established in 2000. The main aim of the Charter is to provide a harmonised platform of space data acquisition and delivery to those affected by natural or man-made disasters.

The members of the charter contribute resources to support the effective implementation of the Charter, contributing to the mitigation of the impacts of disasters on human life and property. The main satellites that can be mobilised within the International Charter include RADARSAT, ERS, ENVISAT, SPOT, IRS, SAC-C, NOAA satellites, LANDSAT, ALOS, DMC satellites, and others. The Charter is manned by a 24-h on-duty operator who receives calls for triggering the charter and can decide immediately.

To further strengthen the use of satellite information, the United Nations General Assembly approved resolution 61/110 of 14 December 2006 which agreed to establish the United Nations Platform for Space-based Information for Disaster Management and Emergency Response – UN-SPIDER. This is a new United Nations programme whose main mission is to "ensure that all countries and international and regional organisations have access to and develop the capacity to use all types of space-based information to support the full disaster management cycle".

A main characteristic of UN-SPIDER is the expressed need to ensure access to space technologies and tools during all phases of a disaster, including the pre-disaster risk identification and assessment phase, which will significantly contribute to an increasing reduction in the loss of lives and property. In the following, some of the outcomes produced by integrating space technologies with other data sources will be presented.

4.3.3.3. Operational planning and communications

In emergency response, a key factor for delivering effective and timely intervention is the ability to communicate. In most cases, when a disaster hits, the local infrastructure gets destroyed. In such cases, satellite communications provide aid agencies with the necessary support enabling them to coordinate their efforts. Satellite communications render both voice and data exchange possible.

The role of the media has become prominent in the way we respond to emergencies. A good media coverage means more visibility and hence more funding for supporting the emergency response. The CNN factor is a real one. The media easily mobilises public awareness-generating resources. Satellites provide the platform for providing a real-time coverage of events in the disaster-hit areas.

4.3.4. Africa-level initiatives to improve food security using space technologies

The African continent stands to gain a lot from the effective use of EO data and space technologies for development planning, poverty reduction, health and education expansion, as well as human capital improvement. Several continental initiatives have been started, including:

4.3.4.1. Climate for Development (ClimDev Africa)

ClimDev is an African programme to integrate Climate Risk Management (CRM) into pertinent policy and development strategy processes across the African continent. It will be implemented under the Joint Secretariat (AUC, ADB and ECA). ClimDev aims to enhance economic growth and progress towards the MDGs by mitigating the vagaries of climate variability and climate change, and to ensure that developmental achievements are climate-resilient in the longer term. The objectives of ClimDev are consistent with all national poverty reduction strategies and also with the AUC-NEPAD plans for improved agriculture and food security, disaster risk reduction and environmental action, and the global programme on the adaptation to climate change agreed upon at the twelfth session of the Conference of the Parties to the Climate Change Convention (COP 12) in Nairobi in November 2006. The recent AU Summit (January 2007) addressed the challenges posed by the potential impacts of climate change on Africa's development, endorsed the ClimDev implementation strategy and requested the AUC to lead the implementation process (which requires the integration of space technology baseline studies).

4.3.4.2. African Monitoring of Environment for Sustainable Development (AMESD)

This is an African programme funded by the European Union which will be implemented under the AUC. The main objective is to improve environmental monitoring for the sustainable management of the environment, contributing to poverty alleviation and sustainable development. The project promotes the use of appropriate Earth observation monitoring systems in support of development policies, and ensures the coherent regional monitoring of the environment to support regional environmental policies and the regional harmonisation of national policies and practices of environmental monitoring and management. The programme will create a platform that facilitates access to Africa-wide environmental information derived from Earth observation technologies.

4.3.4.3. Africa-Wide Vulnerability Analysis and Mapping Network

This is a joint proposal of the African Union Commission and the UN World Food Programme. The main objective is to contribute to the strengthening of the risk management and emergency response capacities of the AU Commission and its member states through enhanced access to relevant and timely information. The aim is to establish and institutionalise, under the umbrella of the AU Commission, an Africa-wide food security and vulnerability monitoring framework that will provide a platform for data and information exchange among AU member states and partner agencies. This project will develop tools and methodologies that generate advanced vulnerability analysis and risk monitoring capabilities, and develop a platform for integrating socio-economic factors with the associated hazards in a GIS environment. This will lead to an early and prompt risk identification capability, which will strengthen the risk management and emergency response capacity of the AUC.

The above initiatives all have a common objective, which is to improve the livelihoods of millions of Africans through sustainable development. The proposed project builds on these projects linking climate change, the monitoring of the environment and socio-economic factors and vulnerability.

4.3.5. Partnership and capacity building

Building an effective food security and vulnerability information management system requires a considerable level of investment in capacity building and developing partnerships. This approach advocates both the development and utilisation of analytical skills in governments, universities, NGOs, the private sector, and civil society.

Developing and implementing effective humanitarian risk assessment and monitoring requires the participation of a wide range of individuals, institutions and partners. In this regard, there is a need to develop a strategy that facilitates effective partnerships between national governments, regional institutions and international bodies on climate risk reduction and management activities. National governments must be made responsible for developing policies and frameworks that facilitate humanitarian risk identification systems, as well as for establishing monitoring systems.

There are several international and regional initiatives dealing with food security monitoring activities that make use of climate information. These include:

- USAID Famine Early Warning System (FEWSNET)
- FAO Global Information and Early Warning System (GIEWS)
- WFP Vulnerability Analysis and Mapping (VAM)
- Global Monitoring for Food Security (GMFS)
- Monitoring Agriculture with Remote Sensing (JRC-MARS FOOD)
- SADC Regional Early Warning System for Food Security (SADC-REWU)
- Regional Centre for Mapping Resources for Development (RCMRD)
- IGAD Climate Prediction and Analysis Center (ICPAC)
- AGRYMET.

Africa is currently heavily dependent on ad hoc early warning and monitoring systems that are run by various international agencies, NGOs and UN agencies. Such systems are often not integrated into national and regional strategies and may not be sustainable. There is a need for promoting an African-led framework ensuring standardised approaches across countries and regions, using the latest space technologies. The many institutional problems which limit the effectiveness of national information and early warning systems can be summarised as follows:

- Lack of appropriate policies and political commitments. Food security may not often be the priority of governments which lack clear policies and regulations on food security and humanitarian interventions;
- Poor integration into government structures. Typically, data collection and reporting functions remain separated among several ministries. Furthermore, remote sensing is considered a luxury and not a vital source of reliable and objective information;
- Separation of data collection activities from the policy-making processes. People who collect and report food security data typically have no independent decision-making responsibility;
- Problematic relationship between donors and governments. Donors provide technical support and funding for national systems, but governments have their own priorities;
- The existing systems generally operate with little coordination among themselves;
- Lack of institutional memory. Systems are rarely institutionalised due to a high staff turnover in governments, donor agencies and NGOs, combined with a "project" approach to early warning; and
- Emergency responses dominate long-term planning. Political enthusiasm and donor support for early warning systems tends to be high in the aftermath of a famine but then steadily ebbs away year by year.

4.3.6. Conclusion

Space-borne systems and related technologies and applications are quickly developing, as new generations of satellites are rapidly becoming available for peaceful uses. Earth observation applications can provide advanced tools that enhance the collection, storage, analysis and integration of spatial/geographic data with related non-geographic information collected from ground surveys and stored in databases. The humanitarian community is currently using various types of EO products to support field-level activities such as agricultural monitoring (crop and pasture), disaster assessments (such as flood extents) and operational planning. The main partners in this area include GMFS, FEWS Net/USGS, FAO, UNOSAT and WFP.

For example, as a result of technological developments and the availability of satellite information, our agricultural monitoring capacity is improving rapidly.

Using the latest satellite technologies, we are now able to more accurately monitor cropped areas, crop development, and estimated yields and production. GMFS is pilot-testing a new methodology for cultivated area estimation using radar data at the beginning of the cultivation season. Through the integrated monitoring of crop models, the water requirement satisfaction index, vegetation vigour (NDVI) and field reports, crop and pasture conditions can be monitored more effectively. Crop failures can be identified several weeks before the end of the agricultural season, thus providing early warning information. It is now possible to estimate the dimension of the expected production by mid-growing season. Depending on the analysis, appropriate decision options can be taken. If the information indicates a crop failure, further assessment will be initiated on whether humanitarian intervention is required, and response planning will be triggered. On the other hand, if the analysis indicates an exceptional surplus harvest, it could be decided to initiate a marketing strategy for avoiding market failure.

In summary, the emerging applications of remote sensing for food security monitoring and risk analysis can offer opportunities for improving the effectiveness of humanitarian decision-making. Furthermore, satellite-based communications will provide opportunities for the improved dissemination and exchange of information.

4.4 Threats, risks and sustainability – answers from space Earth observation data for the management of renewable resources

Rupert Haydn, Stefan Saradeth, Peter Volk, Axel Relin, Thomas Häusler, Sharon Gomez & Tobias Wever

4.4.1. Introduction

The objective of this contribution is to demonstrate and discuss the use of space-borne Earth observation (EO) data related to the assessment of renewable resources. Natural resources include application fields such as agriculture, forestry, the environment, the atmosphere, water, and various deposits. EO technology has now been deployed for nearly three decades for the detection and monitoring of natural resources. Especially during the last decade, new technologies, political framework programmes, and increased public awareness have strengthened and promoted the utilisation of EO data in an increasingly wide range of applications. One of the main drivers for the use of EO technology is the growing number of environmental threats and/or changes.

The utilisation and role of EO technology is demonstrated and discussed here on the basis of case studies, which represent contractual activities relating to water, forests and crops carried out by GAF AG on behalf of the European Space Agency (ESA), the European Union (EU), the World Bank (WB) and others. GAF AG is a Germany-based internationally recognised consulting company providing services based on remote sensing, geo-data and spatial information systems. For more than 20 years, GAF has built a proven track record in more than 100 countries which also includes capacity building, training and technical assistance projects in the natural resources and environment sectors.

4.4.2. The setting

Times have never been more favourable and conditions never better for taking advantage of space-based Earth observation technologies. After a development and demonstration period that lasted for more than three decades, we have finally arrived at a point where several coincident factors are driving the implementation and utilisation of EO on a large scale. Some of these factors can be characterised as follows:

(a) Increasing pressure from, and awareness of, environmental problems,
(b) Mature and, in many cases, readily available technologies (still to be further implemented), and
(c) An open-minded public and/or political framework keen to invest in, and take advantage of, this technology. This also includes the private sector investing in costly space segments and in the development of applications.

The growing need to better understand our environment is not only confined to local and regional dimensions. Many requirements have to be seen in a global context, which requires globally operating information-gathering networks. We are facing a shortfall of non-renewable resources and are endangering our renewable resources by ignoring the rules of their responsible and sustainable handling. The reasons for this may be a lack of knowledge and skills, the short-term pursuit of profits, or even the simple matter of survival in many developing countries. Considering the broad field of causes and effects, it becomes clear that EO can be utilised at various levels to deal with environmental aspects and the preservation of our resources.

It is interesting to note that the initialisation of the first civilian satellite-based EO programme (ERTS-LANDSAT in 1972) was influenced significantly by the geological community in the USA. The first operational users investing in the exploitation of Landsat images were from the oil and gas industry. At the time, all major exploration companies were operating costly image-processing laboratories to take advantage of any information contained in such data that could support their exploration activities. Later on, environmental issues with a strong political component provided the momentum to further develop and implement EO technologies and applications. The problem of forest dieback in the 1980s, related to environmental factors such as "acid rain", was a very hot issue and led to numerous research and development activities worldwide. Today, the problem has been replaced by "global warming". This time, the effects of the increasing number of environmental threats and/or changes are omnipresent and can be seen on a more or less daily basis with EO data and images. In this context, measurements and images taken from space-borne platforms play an important role in fostering public awareness of both the problems and of the technology. Public awareness is an important aspect in creating the political will to support the implementation of EO applications. In this area, the initiatives of Google and Microsoft Network

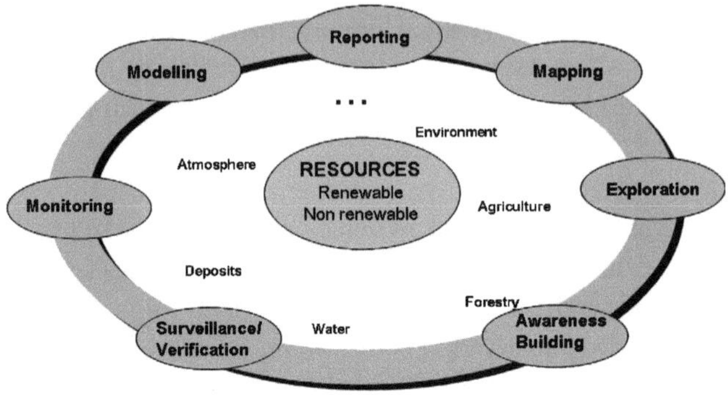

Fig. 5. *The role of EO for the management of renewable and non-renewable resources.*

in offering a global image database to the Internet community have to be appreciated as unique marketing contributions.

In terms of renewable resources, the areas where Earth observation data alone or in conjunction with ground-based and other supplementary information can be applied are extensive. The range may include straightforward mapping activities of a given situation and the mapping of changes over time, which are referred to as monitoring and/or change detection, as well as the analysis of complex situations by means of models, the elaboration of problem mitigation and solving measures and, last but not least, the monitoring of the results of certain actions in terms of control and proof of concept (Figure 5).

4.4.3. Case studies

The following is intended to demonstrate and discuss a few aspects of the use of space-borne EO data related to renewable resources. The examples represent contractual activities related to water, forests, and crops carried out by GAF AG on behalf of the European Space Agency (ESA), the European Union (EU) and the World Bank (WB).

4.4.3.1 Water[84–86]

Water is increasingly becoming a political and strategic resource for many countries on this planet. In semi-arid and arid environments, man and nature have lived together for aeons, relying on shared and sustainable water usage. This has changed with the pressure from a growing human population, the increasing use of drinking water (also per capita) and, in particular, with the use of water for intensive agricultural purposes.

Surface water has become a real scarcity in many areas, and is a political issue of the first order. Water usage in the Middle East is, for instance, a permanent threat to the stability of the region, adding stress to the permanent political tensions. The problem is caused mainly by the fact that national boundaries rarely coincide with natural watershed boundaries.

The possibility of exploiting groundwater resources with limited means was discovered at an early stage, and thus cases of human overexploitation were rare in the past. This changed dramatically in the 20th century due to technological advances. Today, electric and mechanical pumps are in a position to rapidly deplete even very productive aquifers. Legal and regulatory systems are needed if a rapid decline in the sustainability of the water supply is to be avoided.

The Sana'a basin in Yemen is covered by thousands of uncontrolled pumps, which are virtually sucking dry the aquifers that were probably refilled during the last Ice Age when rainfall was more abundant. Some areas in the basin report annual groundwater level declines of up to 5 m. Even huge fossil aquifer systems such as those in central Saudi Arabia or in Libya will collapse, as the temptation to extract even more water is too strong. Alternatives like seawater desalination are more costly and consume other important resources (energy).

Confronted with this situation, it becomes clear that good governance of the water supply and the irrigation sector is a key topic in many countries. The establishment of monitoring systems for the main user groups of water resources is a necessity if sustainable management, or at least controlled depletion, are to be achieved. Earth observation is not the complete solution for building and feeding such systems, but it is a very important constituent element. EO delivers, for example, *timely* and *spatially* dense information on the current utilisation of water for agricultural irrigation, which accounts for up to 60–80% of all water consumed in arid and semi-arid countries.

A good example is the provision of essential spatial hydrological data as input to a water master plan that focuses on improving the sustainability of the Sana'a basin aquifer systems. The project background and recent problems can be described as follows:

The Sana'a Basin (3200 km^2) suffers from a rapid and extreme depletion of groundwater, as it is the only water source for domestic, industrial and agricultural use. The use of water by far exceeds the natural rate of recharge, including irrigation and wastewater return flows, and is inefficient and totally unregulated throughout the basin. About 75% of the population are entirely dependent on agricultural activity, which mainly comprises qat and fruit cultivation as well as livestock raising. The population of 3.0 million grows at a rate of 7% per annum.

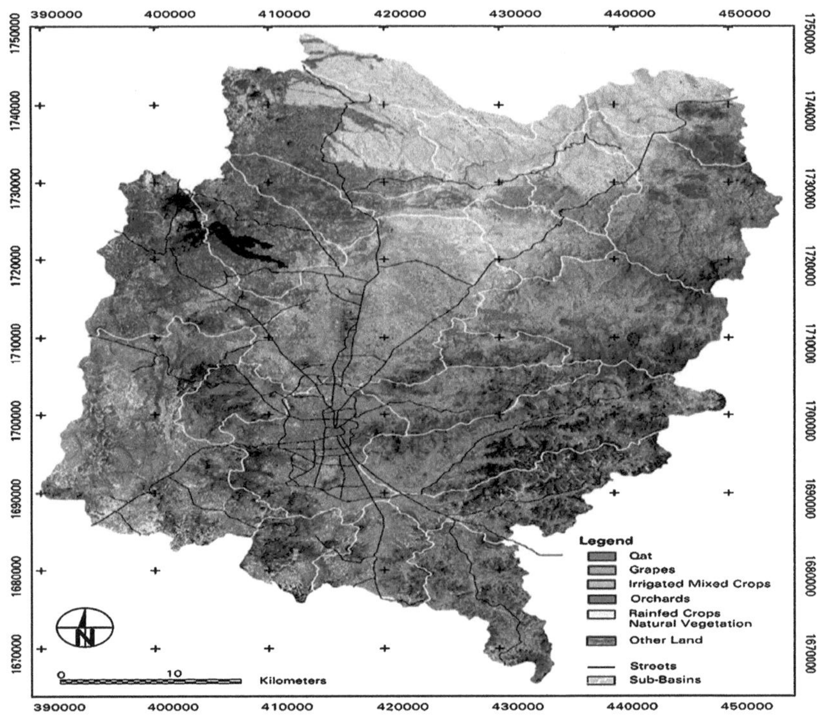

Fig. 6. *EO-derived landuse map of the Sana'a Basin as input for the calculation of the actual evapo-transpiration.*

In order to derive the current water balance in the Sana'a basin, evapo-transpiration (ETa) levels have to be modelled. This process follows a stepwise calculation of the spatial distribution of the actual evapo-transpiration using reference evapo-transpiration and crop evapo-transpiration modelling values. Earth observation data provide vital information for this purpose – namely the land use and the irrigated vegetation density (Figure 6). For the first time, it has become possible to calculate the crop-covered and irrigated areas, cropping patterns and the actual evapo-transpiration levels for the entire Sana'a basin. With these values and other ancillary data, net and total groundwater use was assessed. The estimation of the spatial distribution of groundwater extraction and recharge for the 2004/2005 season was calculated using the mass balance between rainfall and ETa. The net groundwater use in the basin was thereby estimated by calculating the difference between rainfall and ETa, taking into account irrigation efficiency and recharge.

This example highlights how Earth observation data can actually contribute to an efficient management and monitoring plan for the protection of water resources in overexploited regions of our planet even today.

4.4.3.2 Forests as a renewable resource[87,88]

Forests are important ecosystems that have a systemic impact on the biodiversity, the water cycle and the micro- and macroclimates of a region, and are currently under threat worldwide. The most obvious impacts are changes in land use (agriculture, urbanization, etc.), forest fires, and legal and illegal logging in tropical and boreal forests, which all lead to a reduction in the areas covered by forests and in related habitats, as well as to forest fragmentation.

Due to the multi-resource functions of forests, governments must focus on improving their management. Both qualitative and quantitative forest information is needed for the assessment and monitoring of forest resources and for a sustained and productive forest management of the globally available forest resources. The need for improved, spatially accurate and timely data for forest management has been underscored in many national, regional, and international environmental and forest-related policy segments and UN Conventions.

While information on forest area and type was in the past collected on an irregular basis and mostly by means of initiatives of national inventories and monitoring programmes, there is an increasing demand for a more standardised approach to such inventories. In this context, satellite remote sensing can play an important role by providing a data source which has the advantage of being universal. Remotely sensed imagery has the additional advantage of easy integration into Geographical Information Systems (GIS), which can serve as an integrating platform for all relevant data sources. The use of modern Earth observation and GIS techniques for assessing structural diversity and changes to forest ecosystems has some major advantages:

1. The data provide homogeneous and objective information about vegetated surfaces for large areas;
2. The data are inexpensive compared to other methods of data collection;
3. Digital data are an adequate basis for continuous updating as well as for sophisticated analysis procedures together with other spatial information.

The space community in Europe, in an effort to support government requirements for improved information for meeting environmental policy reporting requirements, initiated the Global Monitoring for Environment and Security (GMES) programme in 2003. GMES is a joint initiative of the ESA and the European Commission (EC) and aims at providing operational information to the monitoring and management of the environment as well as for security purposes. The GMES Service Elements (GSE), supported by ESA, focus on the need for policy-relevant end-user services utilising primarily EO sources. The specific GSE related to Forest Monitoring (GSE FM) focuses on the application of EO for improved forest management. The services and products comprise accurate, reliable, timely, and effective information on the state of the global forest systems that support decision-making and improve policies to enable sustainable forest management, compliance with specific protocols and binding conventions, and related user and/or policy-driven activities. The information provided by the monitoring service is delivered in the form of standardised, spatially referenced, high-quality products.

A specific example of the GSE FM's service portfolio is the provision of services/products that are required for the UN Framework Convention on Climate Change (UNFCCC) and the related Kyoto Protocol (KP) reporting. Avoiding deforestation is considered a contribution to the reduction of green house gas (GHG) emissions. However, uncertainties exist about the quantifiable influence of the reduction of forest areas on the carbon balance; for example, there are uncertainties regarding how forest loss and degradation should be defined and how these forest areas can be regularly inventoried (area and associated GHG). Activities related to land use and forestry have a significant impact on the net emissions of carbon. The UNFCCC and Kyoto Protocol impose the obligation on developed countries to report on these activities during the first commitment period 2008–2012. Therefore, these countries need to monitor such activities and measure or estimate their impact on the carbon stock constituted by forests and other vegetation. The GSE FM services provide the information required about these forests for the Kyoto Protocol reference year 1990 and for the subsequent years. The services can be used to generate national summary statistics about totals and changes in terms of the stem volume, biomass, and carbon stock. This includes the error budgets required to define the minimum reliable carbon stock estimate. Forest cover and forest cover change maps, together with appropriate statistical data, are also provided.

The GSE FM services combine EO data with in situ information from a sample-based forest survey. In situ measurements, i.e. tree diameter, height, and species, are used as input for models that provide biomass and carbon estimates. In regions with limited in situ measurements, optical imagery with high resolution can partly replace ground measurements. The principal sources of EO data are the optical imaging instruments. In regions with frequent cloud cover, Synthetic Aperture Radar (SAR) can be applied as an alternative (Figure 7). The specification and acceptability threshold is an accuracy level of 95% (±5%) for the thematic mapping of the forest area and a minimum mapping unit of 0.5 hectares. Examples of European and international bodies that have used the service and have given positive feedback on the suitability of the information for UNFCCC reporting purposes include the Federal

Fig. 7. *Deforestation in French Guiana based on ENVISAT, ASAR and ERS data (source: VTT Technical Research Centre of Finland; Institut Francilien des Geosciences, University of Marne-la-Vallée).*

Ministry of Consumer Protection, Food and Agriculture (BMELV) in Germany, the National Observatory of Athens (NOA) in Greece, the Ministry for the Environment and Territory (MATT) in Italy, the Forest and Landscape Denmark Agency (FLD), and the Indonesian Society for Natural Resources and Environmental Accounting (MASLI).

4.4.3.3 Agriculture[89]

There are many agriculture-related applications where remote sensing plays an important role. These applications range from the collection of "agricultural intelligence" related to crop production and yield forecasts by means of the identification of specific environmental conditions as input for agricultural techniques referred to as "precision farming", to a clearly defined assessment of crop types and acreages on farm levels. To illustrate this, a case representing the latter application has been selected to demonstrate the degree of operational applicability of EO-based services for annually recurring monitoring.

Since 1987, European policy-making has targeted not only relevant R&D projects, but also a range of services in the field of agriculture relating to the implementation of dedicated Agricultural Information Systems. These initiatives make use of in-depth knowledge and long-standing experience and include activities focused on building up excellence in all associated fields. The developed services combine detailed knowledge of the demands of the user community (European Commission, national authorities, agro-industry and farmers) with an in-depth understanding of all related disciplines (GIS, remote sensing, database management and agriculture). They also have the human resources, expertise and motivation at their disposal to ensure the capability to develop and maintain information systems for service providers, agricultural administrations and farmers. This European added-value industry has established an extended project and product portfolio in the field of agriculture and continues to play a leading role in the ongoing development.

4.4.3.4 Subsidy controls[90]

As part of the European Common Agricultural Policy, an "integrated administration and control system" (IACS) was set up for Community subsidies granted under the reform arrangements. The European Agricultural Guidance and Guarantee Fund (EAGGF) annually pays out about 33 billion euros in subsidies to the European farming community. Payments to farmers in all member states are based on standardised annual declarations at the farm level, providing detailed information about declared land parcels, animals, and integrated control procedures. The declarations are compiled in a computerised database in accordance with the specifications laid down by the Joint Research Centre of the EU Commission.

Due to the fact that the payment of subsidies depends on the types of crops cultivated and on the respective areas, remote sensing in combination with GIS functionalities are services heavily used as an objective and uniform method supporting the entire process, from the preparation of the declarations by farmers to the standardised submission to authorities and the verification of the submitted information. The controlling of all declarations submitted is the responsibility of the relevant authorities of the individual member states. The controlling procedures include an accuracy and plausibility check of all declarations submitted. Per definition of the Community, at least 5% of the declarations have to be verified in situ. The field checks are significantly facilitated by the use of remote sensing.

On the basis of competitive tendering procedures, about 20 highly specialised remote sensing service companies are selected annually by the authorities of the

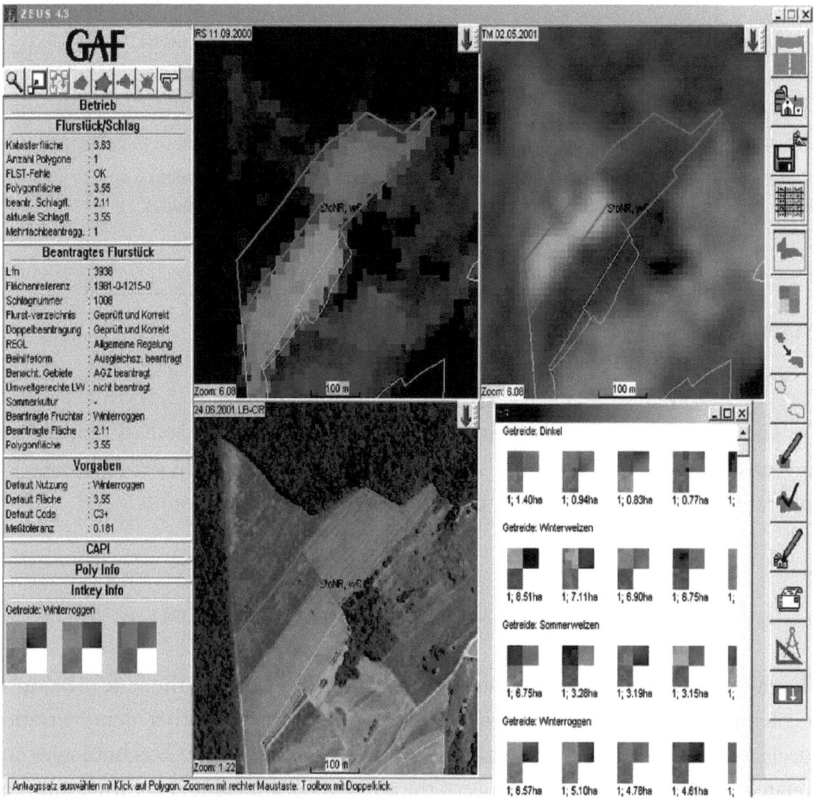

Fig. 8. *Interactive software ZEUS for the control of subsidy declarations with multitemporal and multisensory EO data.*

member states to carry out an assessment using remote sensing data. This involves a range of sources in order to achieve as many snapshots as possible (2–3) of the phenology of the crops during the growing season and to carry out area measurements in accordance with the EU accuracy requirements. Semi-automated processing systems involving human interaction at certain levels have been developed and are being applied in order to avoid erroneous results and to optimise processes. Without such dedicated software tools (Figure 8), it would not be possible to manage the annually recurring workloads in terms of quantity and quality. Such a workload consists typically of 1,000 farms with 30,000 fields that have to be inspected and documented by a single contractor within 3–4 weeks.

This approach reduces the number of farms to be controlled in situ by focusing on problematic declarations and parcels. Furthermore, it facilitates the remaining checks by providing tabular and graphic products.

Number of declarations:	8.9 million euros
To be controlled (min. requirement):	5%
Number of farms controlled:	587,000 = 6.6%
Controlled with remote sensing:	286,000
Number of participating countries:	24
Financial volume of subsidies:	33 billion euros
Costs of control with remote sensing:	50 million euros
Contractors – private companies:	31
Governmental organisations:	12

Fig. 9. *Key figures on 2006 subsidy controls with remote sensing.*

These control activities constitute the largest European EO activity tendered on a yearly basis (Figure 9). It has steadily grown since 1992 and demonstrates the operational potential of EO to provide reliable and robust results.

4.4.4. How to make EO services operational

Today, satellite-based Earth observation has gone well beyond the state of being an evolving technology that has to be justified by pilot and other demonstration projects. The obstacles to the widespread implementation of EO technologies and related applications consist mainly of the costs of a sustainable space infrastructure and a lack of effective governance ensuring that the opportunities and benefits of EO be fully exploited. With regard to governance, fewer efforts were devoted to the development of applications in the past; in other words, the development of EO was more technology-driven than application-driven.

This may be one of the reasons why today, we are able to rely on several dozen Earth observation satellites currently in orbit. Data from these systems are being made available on both a commercial and a not-for-profit basis, and they form the backbone of the current applications and developments.

After recognising that public authorities constitute the majority of EO users, the joint ESA/EU GMES initiative was created. GMES is supplied with adequate funding and is strongly supportive of applications and users. The initiative aims to develop an end-to-end system consisting of a broad user community and a technical and commercial infrastructure capable of supporting these users in supplying their services to the public.

However, if we look at the realisation of such a programme over the longer term, there are still a lot of open issues beyond the implementation and validation phase related to the financing of services as well as to the contractual allocation of work in a highly competitive environment. The players offering EO services are found not only in the private sector, but also consist of governmental and semi-governmental organisations and other national and international establishments partly or fully financed by the public. This fragmented scenario requires a fair mechanism that allows the private sector to play a role vis-à-vis governmental and publicly supported organisations that may be both users and service providers at the same time.

Concerning the financial aspects, a differentiation must be made between funding and budgeting. Until GMES becomes operational, its funding is assured, including the space segment. As regards public organisations as users, sufficient budgets have to be available in the future to guarantee the provision of EO services beyond the abovementioned developmental phase. Drawing on past experiences, the implementation of a suitable budgetary policy should be an integral part of such an initiative in order to enable the seamless transition from development to operation.

[84] European Space Agency. Water Resources Management in Africa (WSSD Follow-Up). ESA/PB-EO(2003)24, 7 Feb. 2003.

[85] Saradeth, Stefan and Weissmann, Thomas. "AQUIFER: Remote Sensing as Support for the Management of Internationally Shared Transboundary Aquifers in Africa." The Future of Drylands: Proceedings of the International Scientific Conference on Desertification and Drylands Research, Tunis, Tunisia, 19–21 June 2006. eds. Lee, Cathy and Schaaf, Thomas. Dordrecht, Netherlands: Springer.

[86] Food and Agriculture Organization. Crop Evapotranspiration: Guidelines for Computing Crop Water Requirements. FAO Irrigation and Drainage Paper 56. Rome: FAO, 1998.

[87] Global Monitoring for Environment and Security. "GSE Forest Monitoring." 1 Dec. 2007. http://www.gmes-forest.info/

[88] European Space Agency. "GMES Services Element (GSE)." 1 Dec. 2007. www.esa.int/esaLP/SEM36I2IU7E_LPgmes_0.html

[89] Relin, Axel, Krause, Arno and Zeug, Gunter. IACS-GIS 2005: Demands and Solutions. Proceedings of the 4th European Conference for Information Technology in Agriculture (EFITA), Debrecen, Hungary, 2003.

[90] Commission Regulation (EEC) No. 796/2004 of 21 April 2004 laying down detailed rules for the implementation of the cross-compliance, modulation and integrated administration and control system provided for in Council Regulation (EC) No. 1782/2003 establishing common rules for direct support schemes under the common agricultural policy and establishing certain support schemes for farmers. See: Official Journal of the European Union L 141/18, 30 Apr. 2004.

CHAPTER 5

ENERGY

5.1 Overview

The context: coping with the increased energy needs caused by a growing world population and increasing industrialisation is one of the mega topics of 21st-century international politics. Energy policy has to follow three inherently contradicting goals: the security of supply, efficiency and cost effectiveness as well as environmental compatibility. If a balance cannot be reached globally or on the regional or national level, severe conflicts will arise. Technology developments will therefore have a decisive effect on the evolution of this crucial field.

The contribution of space: space provides tools for strategic decision-making in the energy sector. Space applications support the planning and monitoring of pipelines and the operational needs of the energy sector (like weather forecasting, seismic testing, security supervision, etc.). Remote sensing is a key tool for assessing, planning and monitoring exploration surveys and is inter alia used for hydrocarbon exploration and geophysical exploration. Satellite applications can also contribute to the problems of finding the right locations for power plants, and monitor renewable energy production. Europe has various satellite systems which can respond to the respective needs and support the European energy policies at the supra-national as well as the national level. Space might also contribute to solving mankind's energy problems in the more distant future. This comprises concept studies for satellites in geostationary orbit relaying solar power produced there to the Earth, or the extraction of ^{3}He from the Moon used for fusion on Earth.

The papers: the three papers of this section provide not only a general overview of the field but also range from concrete applications to the most visionary approaches.

In "Achieving Sustainable and Secure Energy Supplies in a Carbon-Constrained World", Galina Zlatareva, an energy consultant with Louys Consulting, details the setting for energy policy as a global and European topic of high relevance. She investigates the need for a growing share of renewable energy in order to satisfy Europe's needs and become less detrimental to the environment at the same time. She also indicates where space has to make an impact either as an instrument for securing the energy supply or in terms of visionary plans to replace the current energy sources.

In "Remote Sensing for Sustainable Hydrocarbon Exploration and Reservoir Monitoring", Andreas Laake from WesternGeco, the world's leading geophysical services company providing a wide range of services to the oil and gas industry, provides in greater detail and with numerous illustrative examples an overview of space applications' services to the energy sector. In this context, he reflects on the wealth of methods as well as products in this field, which he divides into prospect identification, geophysical surveying, and reservoir management.

In "Sustainable Energy Supply on Earth – Can Space Contribute?", Wolfgang Seboldt from the German Aerospace Center DLR, the major engineering research centre in Germany which deals not only with aerospace but also with energy research, opens the perspective to visionary ideas solving humankind's energy problems through space-based energy production. In addition, he shows how spin-off effects of space research have promoted the use of regenerative energy resources on Earth, for instance by solar cells which were originally developed for space.

5.2 Achieving sustainable and secure energy supplies in a carbon-constrained world

Galina Zlatareva

5.2.1. Introduction

Energy is vital for the economic development of the world and is becoming a substantial part of modern societies' everyday life. However, the days of cheap energy seem to be over. Today, the world faces challenges such as climate change, increasing import dependence and higher energy prices. The remaining hydrocarbon resources are highly concentrated in politically unstable regions, thus enhancing the vulnerability of the Western world and creating a strong leverage for the producer countries to use their resources as a political weapon. According to the World Energy Outlook prepared by the International Energy Agency, the world's primary energy demand is set to rise by 59% from now until 2030, with some 85% of the increasing demand coming from fossil fuels and two-thirds from the developing world, especially China and India. However, if actions are taken, these predictions could be altered.

Although climate change issues are becoming more and more urgent, the negotiations at the recent World Climate Conference in Bali in 2007 failed to yield an international agreement for binding measures to reduce CO_2 emissions. It seems as if the political will for commitments in this field is still not strong enough.

5.2.2. Growing demand and declining resources

The world is becoming increasingly dependent on imported hydrocarbons that carry political and economic risks while increasing pressure on the global energy resources. The International Energy Agency (IEA) expects the global demand for oil to grow by 41% by 2030.[91] Furthermore, the IEA stated that "the ability and willingness of major oil and gas producers to step up investment in order to meet rising global demand are particularly uncertain". As long as there is an increase in consumption, rising prices and a lack of diversification in the energy mix, the risk of

Tab. 1: *Energy dependency of the EU (source: IAEA).*

	1994	1995	1996	1997	1998	1999	2000	2001	2002	2003	2004	2005
EU (27 countries)	42.8	43.3	43.9	44.8	46.0	45.0	46.6	47.3[a]	47.4[a]	48.8	50.1[a]	52.3[a]
EU (25 countries)	43.1	43.5	44.1	45.1	46.4	45.5	47.2	47.7[a]	47.9[a]	49.3	50.5[a]	52.9[a]
EU (15 countries)	46.1	46.5	46.7	47.7	48.8	47.6	49.5	50.2[a]	50.2[a]	51.7	52.9[a]	55.4[a]
Belgium	79.4	80.5	80.5	78.4	81.3	76.6	77.8	80.0	76.4	78.8	78.9	78.2
Bulgaria	55.6	57.2	56.0	52.6	49.8	48.9	46.5	46.2	46.7	46.9	48.0	47.1
Czech Republic	19.2	20.6	24.3	24.3	25.5	25.1	23.1	25.7	26.3	24.9	24.5	27.4
Denmark	27.3	34.5	23.2	17.3	7.0	−15.2	−33.7	−26.6	−41.0	−31.6	−47.5	−51.6
Germany	56.7	57.3	59.1	59.9	61.4	59.5	59.8	60.9	60.0	60.7	61.2[a]	61.6[a]
Estonia	38.6	35.7	33.1	33.3	37.5	36.6	30.8	31.0	28.3	26.0	28.1	25.8
Ireland	65.2	69.4	71.1	77.1	81.1	84.3	84.5	90.3	89.0	89.3	86.6	89.5
Greece	58.6	65.7	66.0	66.8	69.5	66.1	69.3	68.8	70.6	67.3	72.6	68.5
Spain	68.3	71.6	70.1	71.8	74.3	76.5	76.5	74.3	78.3	76.5	77.3	81.2
France	47.6	47.9	48.4	48.9	51.1	51.2	50.9	50.4	50.8	50.5	50.7	51.6
Italy	81.3	82.3	82.3	81.0	82.1	82.8	87.3	84.0	86.5	83.8	84.6	84.4
Cyprus	97.7	99.0	98.3	98.3	96.8	100.5	98.8	96.1	100.5	96.1	96.5	100.7
Latvia	71.4	68.4	71.4	57.5	57.5	53.5	57.0	56.7	54.9	57.8	63.7	56.0
Lithuania	65.3	64.0	53.6	56.5	51.0	54.4	60.4	47.5	42.6	45.0	47.8	58.4

Luxembourg	98.7	97.7	99.3	98.4	99.5	97.3	99.8	97.9	99.0	98.7	98.2	98.0
Hungary	47.3	48.8	52.6	52.8	56.1	54.7	56.0	54.5	57.0	60.7	60.5	62.9
Malta	100.0	100.0	100.0	100.0	100.0	100.0	100.0	100.0	100.0	100.0	100.0	100.0
Netherlands	21.0	19.3	16.0	26.0	26.9	29.7	38.5	34.1[a]	33.9[a]	37.7	31.0	37.8
Austria	65.1	66.5	69.6	66.9	70.3	65.7	65.6	64.1	67.6	70.0	70.5	71.8
Poland	0.0	–0.2	5.2	6.3	8.3	10.1	11.1	10.3	11.2	13.0	14.7	18.0
Portugal	81.2	85.9	80.5	84.0	83.2	87.7	84.7	85.4	84.1	85.1	83.6	88.2
Romania	25.9	30.9	31.0	32.6	28.2	21.6	21.8	25.7	23.7	25.3	30.2	27.4
Slovenia	48.2	50.1	54.5	55.2	52.2	55.3	52.5	50.2	50.1	53.3	52.1	52.2
Slovakia	70.2	70.6	74.9	74.5	71.5	67.4	66.5	63.3	64.9	65.5	68.7	64.6
Finland	66.0	53.1	55.4	55.9	54.3	51.6	55.9	56.1	52.5	59.1	55.0	54.7
Sweden	40.0	37.5	40.2	38.2	37.8	34.9	39.0	36.3	37.8	43.6	37.0	37.2
United Kingdom	–13.6	–16.3	–14.4	–15.3	–15.6	–20.3	–16.7	–9.1	–12.2	–6.2	4.9	13.9
Croatia	42.9	41.0	44.4	47.5	49.5	54.6	53.3	52.2	60.1	56.3	57.4	58.6
Turkey	55.5	59.1	60.5	60.0	59.9	60.8	65.4	64.3	67.2	70.9	70.4	71.9
Iceland	35.0	33.8	34.4	33.9	33.9	30.9	31.2	27.7	27.9	27.1	30.0	28.8
Norway	–604.7	–642.9	–759.7	–736.5	–682.4	–659.2	–735.8	–732.3	–834.3	–744.6	–728.9	–609.1

[a] Provisional value

a supply failure will continue to grow. For an overview of the EU's energy dependence, see Table 1.

The recent increases in energy world market prices, with costs of over 130 dollars per barrel of oil with predictions to reach 200 dollars, combined with the alarming climate change issue, have highlighted the urgency of the challenge to find a solution for securing future energy supplies and finding new alternatives. Furthermore, the available resources have become even more concentrated in politically unstable regions with the gas production decline in Norway, which used to be a secure and reliable source of energy supplies.

5.2.2.1. Oil and gas as a political weapon

The world is getting more and more dependent on fossil fuels, with demand rising every day regardless of the increasing prices on the market. This situation, which at first glance seems to be just an economic problem, has strong political implications. The world's available natural resources are unevenly concentrated in politically unstable regions, making the consumer countries an easy target of political pressure. This was the case during the first oil crisis in 1973, when oil was used as a political weapon against the Western countries. The oil-producing Arab countries reduced production and imposed an embargo on the USA, Holland and South Africa, aiming to change their policies towards Israel.[92]

In 1979 and 1980, during the second oil revolution, the vulnerability of the Western world was again misused for political reasons. Today, rising prices and the speculations of OPEC about the oil-peak are creating heavy pressure and uncertainty regarding the security of supply. The risk of dependence on the natural supply of these politically unstable countries also carries the threat of long-term interruptions of deliveries in the case of conflict.

Despite the efforts to introduce alternative energy sources, fossil fuels still have to cover 90% of the worldwide energy demand, with oil covering 37%, which is why oil will probably remain the world's most widely used energy source.[93] The situation in the gas sector is similar regarding the concentration of resources. While 65% of the world's oil reserves and 35% of the gas reserves are concentrated in the Middle East, 38% of the gas reserves are located in Russia.

With the acquisition of 51% of Gazprom's shares by the state, the company has become the strongest political weapon and instrument of the Kremlin to revive its power in the international arena. Nevertheless, Gazprom has recently been behaving like an unreliable monopolist guaranteeing neither future supplies nor stable prices. Moreover, Russia's gas production seems to be declining due to long-term underinvestment in infrastructure. If Russia wants to keep its position as a main supplier of natural gas in Europe and at the same time cover the rising domestic energy demand, there is an urgent need for investments in infrastructure and technology, especially since Russia is facing gas and oil production shortages. The exploration of the two fields – Yamal and Stockman – which could cover the rising demand and compensate the decline in the production of energy exports will require huge investments, most of which will probably have to come from foreign companies. However, the country's legislative basis for foreign investments creates uncertainty and obstacles rather than incentives for foreign investors. Thus, Russia's reliability in terms of its ability to supply the needed amount of gas, and of delivering it without interruptions, is becoming unsure.

The first conflict regarding energy supplies from Russia was the *Russia–Ukraine crisis* in 2006. The core of the dispute was the unilateral decision of Gazprom to enforce new commercial prices on natural gas in the Ukraine. The dimension of this dispute is quite disturbing, taking into account the fact that 80% of Russia's gas supplies come from pipelines running through the Ukraine. Even though a complex agreement on the new price for gas deliveries was reached, the challenge of the EU's reliance on energy resources from unstable or politically tense regions remains on the table. As a consequence of the lessons learned, an EU Gas Coordination Group was created on 4 January 2006. This group of representatives from member states, industry and consumers was set up on the basis of EU Council Directive 2004/67/EC of 26 April 2004 concerning measures to safeguard the security of natural gas supply.

The problem with the Russian supply was further sharpened by the *Russian – Belarus conflict* in January 2007. The closure of the Druzhba pipeline (the main conduit of Russian oil to Europe) for 3 days left parts of Europe without its needed supply. Thus, the fear of Russia using its 'gas wealth' as a geopolitical weapon has put energy issues high on the EU's political agenda. The political strain over strategic issues with Russia has cast clouds on the relationship, as is reflected in Brussels' provoking a debate on the need of a common European energy foreign policy towards the supplier-countries: "... At the highest level, the aim of implementing the common internal energy policy and setting an external energy policy of the EU has been set".[94] Kremlin's strategy has shown that in the face of an increasingly energy-vulnerable Europe, Russia's gas supplies cannot be taken for granted anymore.

These events provoked lively discussions about seeking further possibilities for diversifying Europe's energy sources by intensifying relations with alternative suppliers or endeavouring to change Europe's energy mix by making wider use of renewable energies.

The anxiety of Europe was further heightened as a result of the recent diplomatic actions of Russia which aimed at diminishing the possibilities of Europe to diversify its gas suppliers and transit routes. The Nabucco pipeline project is a crucial contribution to Europe's diversification endeavour. Transporting Caspian gas through Austria, Hungary, Romania, Bulgaria and Turkey and bypassing Russia, the Nabucco pipeline is regarded as a priority objective of the EU: "The development of major international pipelines to deliver oil from the Caspian region and Central Asia to the EU is vital".[95]

The "great game"[96] for the Caspian resources seems to have been re-opened, as the EU and the USA backed the project's realisation which had been questioned following Russia's undermining actions. After signing a long-term contract with Turkmenistan and Kazakhstan to build a new pipeline, which would carry Central Asian gas into the Russian Federation instead of establishing a direct connection to the EU, Nabucco's commercial viability has become questionable, as Turkmenistan may not be able to produce enough gas for both projects.

5.2.2.2. The Gas-OPEC perspective

In the context of projections for an increase in the world's gas consumption, Russia is trying to spread its monopoly structure to a wider area. On 9 April 2007, the Gas Exporting Countries Forum, which controls 73% of the world's proven gas reserves and 43% of the current gas production, announced the idea of building a Gas-OPEC for cooperation in production and price-setting mechanisms.[97] The negotiations between Russia, Algeria, Iran and Venezuela on an initiative to create an international gas cartel are creating additional pressure on the gas market, although there are differences between the two commodities, thus preventing the cartelisation of the gas sector at least in the short term. Natural gas is sold regionally through pipelines on the basis of long-term bilateral contracts, while oil is a global commodity traded on the commodity market. If this obstacle is to be overcome, major investments are still needed to make the LNG (Liquefied Natural Gas) technology economically reasonable compared to the pipeline option. Furthermore, posing another obstacle to the establishment of a Gas-OPEC, Vladimir Milov, President of the Institute of Energy Policy in Moscow, pointed out the fact that Iran and Venezuela are hardly net exporters and that the gas markets are still strongly regionally fragmented. However, closer cooperation between these politically unstable countries towards the centralisation of decision-making and a production and pricing regime based on the OPEC-model is a real threat to the Western consumer countries.

5.2.2.3. The emerging superpowers (China and India) – increasing the tension on the energy market or a chance to develop a new approach to energy policy?

The Asian market, with its fast-growing economy and high consumer potential, will have serious impacts on the future constellation of the energy market by creating a high increase in demand. The rising number of buyers and the restricted amount of available natural resources will create competition with consequences for the prices of the remaining commodities. Moreover, the growing Asian market will have environmental repercussions and will impact the world's energy security.

The threat to global energy security consists in a rising demand for additional energy resources, and one should keep in mind that not only the USA and Europe, but also the newly rising powers will compete on the energy market. The economies of the ASEAN countries are growing fast, as is their energy consumption. In the period between 2000 and 2004, China was responsible for 30% of the rise in the global energy consumption, and demand will continue to grow.[98] The country's oil and coal resources will not be able to meet the growing demand in the future due to growth in the high-energy consuming branches of the economy.

In this case, the temptation to continue to use the country's available coal resources to satisfy demand is huge, unless new policies were implemented that would ensure the support needed for the introduction of renewable energies. Due to the favourable natural conditions for the development of wind energy in the region, there are high renewable energy potentials. Despite the efforts of China and other industrialising countries to increase the role of renewable energy sources in their energy mix, the results are still minor.

Thus far, India has the lead with over 900 megawatts of wind power in place,[99] but to ensure sustainable development and to enhance investment in this sector, a long-term policy strategy is needed that would ensure support and tax incentives. Unlike India, China has not yet established a solid legal basis for a sustained wind power industry, although the country has a clear potential to become a wind superpower with its huge wind resources in several regions, including Inner Mongolia (an autonomous Chinese territory with the potential of becoming the 'Saudi Arabia of wind power.' In case it is well explored, China's wind potential is estimated to exceed the country's total current generating capacity. Although it is ranked fifth in total installed capacity (6,050 megawatts), an estimated one-fourth of this capacity remains unconnected to the grid due to planning problems.[100]

The Western world has the moral responsibility and duty to help emerging economies in making a leap in evolution, by sharing with them some of the best practices learned in its renewable energies sector. The developing world could benefit greatly from further growth in the renewable energies industry. In this regard, the governments of the leading countries in the renewable energy sector (such as Germany and Spain) could introduce policies supporting the dissemination of efficient energy practices.

China's economic growth and energy demand has security, environmental and social implications. Introducing renewable energies could encompass solutions for all these aspects by securing a supply of clean energy, contributing to the fight against climate change and creating many new jobs in the sector.

In this respect, a stable legislative basis is needed to support the future development of the sector and the current subsidies for fossil fuels should be redistributed to support the Carbon Capture and Storage (CCS) technology to ensure the clean use of natural recourses.

5.2.3. Sustainability

According to the European Environment Agency, energy accounts for 80% of all greenhouse gas (GHG) emissions in the EU, which makes them the main source of climate change and most air pollution. Attaining the sustainability of our energy supply will mean that mankind's expanding needs will be met without environmental detriment. The sustainability of the future energy supply, based on renewable resources, is a long-term strategic objective regarding the limited fossil resources and the urgent need to avoid emissions and toxic waste.

5.2.3.1. Renewable energy sources and CO_2 reduction alternatives

Developing the renewable energy sector is essential if we are to achieve a sustainable and secure energy supply for the future; however, investments in the R&D sector are a precondition for achieving the widespread use of the technology by reducing costs. Boosting investment, in particular in energy efficiency and renewable energy, should create jobs, promote innovation and is the necessary step towards achieving sustainable and low-carbon energy supplies.

Energy efficiency is another crucial point which should not be neglected when speaking about establishing a comprehensive framework for an energy policy facing global challenges. The impact of the potential energy saving and CO_2

reduction is enormous – improved energy efficiency alone could cut around 20% of current global CO_2 emissions, according to the IEA.[101]

While the USA, the world's biggest emitter of CO_2, is not yet ready for a binding commitment and a clear target, the EU is clearly declaring its determination to lead the global fight against climate change. In light of the new reform plans in the field of EU energy policy as specified in the comprehensive document "An Energy Policy for Europe"[102] which sets the new framework for the European Energy Policy, the Commission launched the agenda "Limiting Climate Change to 2°C – Policy Options for the EU and the World for 2020 and Beyond".[103] A central part of this document is the EU's objective in international negotiations of a 30% reduction in greenhouse gas emissions for the developed countries by 2020, compared to 1990. In addition, by 2050, global GHG emissions must be reduced by up to 50%, compared to 1990, implying reductions in the industrialised countries of 60–80%. Regardless of the success of international negotiations in this field, the EU has made a commitment to achieve, in any event, at least a 20% reduction of greenhouse gases by 2020, also compared to 1990.[104] For a detailed overview of the EU's renewable energy shares in gross inland consumption as well as for the EU's total GHG emissions, see Tables 2 and 3.

However, since the current situation could be described as a labyrinth of energy and climate change problems, controversial options for sustainable energy – like nuclear energy and biofuels – are at the top of the political agenda.

In a time when the world is looking for sustainable and environmentally friendly solutions for the future, energy supplies which do not require long-term subsidising – which would interfere in the economy and create additional costs for the industries to adapt to the new regulations – do seem attractive. Thus, the nuclear energy lobby is raising the issue of reconsidering nuclear energy as a low-carbon energy source.

5.2.3.2. The nuclear energy option

Nuclear energy is part of the political debate with regard to the strong need for diversification of the energy mix. This option has a unique potential as a large-scale sustainable energy source. It is a low-carbon energy source and for this reason, the issue of intensifying nuclear energy is re-appearing at the top of the political debates. However, the question of nuclear waste disposal remains yet to be solved. Furthermore, it is also questionable to what extent the use of this energy source will contribute to the security of the supply, taking into account that uranium demand is also met largely by imports. Figures 1 and 2 give additional information about uranium production and supply forecasts.

Tab. 2: *Renewable energies in gross inland energy consumption (source: Eurostat).*

	1994	1995	1996	1997	1998	1999	2000	2001	2002	2003	2004	2005
JJD	Renewable energies											
EU (27 countries)	5.1	5.1	5.1	5.4	5.5	5.6	5.8	5.9[a]	5.8[a]	6.0	6.4[a]	6.7[a]
EU (25 countries)	5.2	5.2	5.1	5.3	5.4	5.5	5.7	5.8[a]	5.7[a]	5.9	6.3[a]	6.5[a]
EU (15 countries)	5.3	5.3	5.3	5.5	5.6	5.7	5.9	6.0[a]	5.8[a]	6.0	6.4[a]	6.7[a]
Belgium	1.2	1.4	1.3	1.2	1.3	1.3	1.3	1.5	1.5	1.9	2.1	3.5
Bulgaria	1.1	1.6	2.0	2.4	3.4	3.5	4.2	3.6	4.4	4.9	5.2	5.6
Czech Republic	1.8	1.5	1.4	1.6	1.6	1.9	1.5	1.7	2.1	3.5	4.0	4.1
Denmark	7.2	7.6	7.2	8.3	8.7	9.6	10.8	11.4	12.4	13.5	15.1	16.2
Germany	1.9	1.9	1.9	2.2	2.4	2.6	2.8	3.0	3.4	3.6	4.0[a]	4.8[a]
Estonia	9.3	8.8	10.1	10.3	9.5	10.2	10.8	10.4	10.3	9.5	10.6	11.2
Ireland	1.7	1.5	1.6	1.6	1.9	1.8	1.8	1.7	1.9	1.8	2.1	2.7
Greece	5.1	5.3	5.4	5.2	4.9	5.3	5.0	4.5	4.7	5.1	5.1	5.2
Spain	6.2	5.5	7.0	6.4	6.1	5.2	5.7	6.6	5.5	7.0	6.5	6.1
France	7.7	7.5	7.0	6.8	6.7	6.9	7.0	7.1	6.4	6.4	6.3	6.0
Italy	5.4	4.8	5.2	5.3	5.5	5.8	5.2	5.5	5.3	5.9	6.8	6.5
Cyprus	0.6	2.1	2.0	2.0	1.9	1.9	1.8	1.8	1.8	1.7	2.0	2.0
Latvia	23.8	27.5	27.3	30.5	34.5	34.4	34.3	34.1	34.5	33.1	36.0	36.3

Lithuania	5.9	5.7	5.7	6.1	6.5	7.9	9.2	8.4	8.1	7.9	8.1	8.8
Luxembourg	1.4	1.4	1.2	1.4	1.5	1.3	1.6	1.3	1.4	1.4	1.6	1.6
Hungary	2.3	2.4	1.9	2.0	1.9	1.9	2.1	1.9	3.4	3.4	3.7	4.2
Malta												
Netherlands	1.6	1.6	1.8	2.1	2.2	2.3	2.4	2.4[a]	2.6[a]	2.6	2.9	3.5
Austria	21.9	22.0	20.6	21.1	20.8	22.5	23.2	22.2	22.2	19.3	20.8	20.5
Poland	4.0	3.9	3.7	3.8	4.1	4.0	4.2	4.5	4.6	4.5	4.7	4.8
Portugal	17.8	16.3	18.8	17.5	16.3	13.7	15.4	15.7	14	17.1	14.9	13.4
Romania	5.2	5.9	8.0	10.7	11.2	11.9	10.9	9.3	9.8	10.0	11.7	12.8
Slovenia	10.3	9.4	9.8	8.1	8.7	8.7	12.3	11.5	10.9	10.3	11.6	10.6
Slovakia	3.2	2.8	2.1	2.5	2.5	2.7	2.8	3.9	3.7	3.3	3.9	4.3
Finland	19.0	21.2	19.9	20.6	21.8	22.1	23.9	22.4	21.9	20.9	23.0	23.2
Sweden	23.5	26.0	23.4	27.4	28	27.0	31.4	28.3	26.3	25.3	25.8	29.8
United Kingdom	0.9	0.9	0.8	0.9	1.0	1.1	1.1	1.1	1.2	1.3	1.5	1.7
Croatia	9.9	10.1	13.9	11.0	10.5	11.3	11.2	10.7	9.2	9.1	11.0	10.1
Turkey	18.3	17.4	16.6	15.8	15.9	15.1	13.1	13.2	13.4	12.7	13.2	11.9
Iceland	65.8	67.5	65.5	66.8	67.6	71.3	71.4	73.2	72.8	72.8	72.3	73.0
Norway	45.9	48.9	43.3	43.7	44.0	44.8	51.0	44.1	51.7	38.3	37.7	40.4

Tab. 3: *Total greenhouse gas emissions (source: Eurostat). Index base year = 100.*

Geo / Time	1995	1996	1997	1998	1999	2000	2001	2002	2003	2004	2005	Target
EU(27 countries)	93.4	95.5	93.6	92.7	90.7	90.7	91.7	90.9	92.8	92.8	92.1	–
EU(25 countries)	92.1	94.2	92.8	92.3	90.8	90.7	91.7	90.9	92.4	92.7	–	–
EU(15 countries)	96.9	98.9	97.4	97.8	96.3	96.6	97.7	97.1	98.7	98.8	98.0	92.0
Belgium	103.6	106.4	100.6	104.0	100.1	100.4	100.0	98.8	100.8	100.5	97.9	92.5
Bulgaria	65.6	64.1	61.7	55.0	51.0	50.7	50.9	48.7	52.7	52.2	52.8	92.0
Czech Republic	78.7	82.2	78.5	76.5	72.3	75.9	76.1	73.4	75.2	75.0	74.2	92.0
Denmark	110.0	129.4	115.4	109.6	105.0	98.4	100.6	99.4	107.0	98.4	92.2	79.0
Germany	88.9	90.5	87.5	85.3	82.8	82.7	84.1	82.6	83.6	83.2	81.3	79.0
Estonia	53.8	56.1	55.0	49.3	45.2	45.9	46.1	45.0	50.1	49.2	48.0	92.0
Ireland	106.4	110.2	113.1	118.2	120.7	123.9	127.2	123.7	123.4	123.1	125.4	113.0
Greece	101.9	105.1	109.6	114.2	114.1	118.6	120.0	119.8	123.6	123.9	125.4	125.0
Spain	110.0	107.4	114.6	118.2	127.9	132.8	133.0	139.0	141.5	146.9	152.3	115.0
France	99.1	102.0	100.7	103.1	100.2	99.3	99.7	98.2	98.6	98.6	98.1	100.0
Italy	102.5	101.2	102.3	104.4	105.6	106.6	107.8	107.8	110.8	111.7	112.1	93.5
Cyprus	119.5	125.1	126.5	135.0	137.7	144.7	143.7	152.0	158.4	163.4	163.7	–
Latvia	48.2	48.5	46.3	44.3	41.3	38.8	41.5	41.2	41.7	41.4	42.0	92.0
Lithuania	45.3	47.4	45.3	48.1	42.0	38.9	40.7	41.2	41.7	43.8	46.9	92.0

Luxembourg	77.0[a]	77.6	72.6	66.2	70.9	75.2	77.5	84.9	88.6	100.8	100.4	72.0
Hungary	65.9	67.7	66.1	65.6	65.8	64.3	65.8	64.0	66.8	54.7	65.5	94.0
Malta	122.4	123.4	119.9	121.8	125.9	129.0	118.4	139.8	140.0	145.9	154.8	–
Netherlands	104.9	108.6	105.3	106.1	100.4	99.9	100.8	100.5	101.1	101.8	98.9	94.0
Austria	101.7	105.9	105.4	104.6	102.3	102.7	107.7	109.8	117.7	115.5	118.1	87.0
Poland	77.2	80.8	78.7	73.7	71.4	69.0	68.5	66.0	68.4	67.6	68.0	94.0
Portugal	116.7	112.7	118.5	126.6	138.8	135.0	137.0	144.5	136.1	138.9	140.4	127.0
Romania	56.2	68.2	61.2	54.6	48.0	49.1	50.6	53.3	55.8	56.7	54.4	92.0
Slovenia	91.4	94.6	96.2	95.0	91.6	92.6	97.1	98.1	96.8	98.4	100.4	92.0
Slovakia	72.3	73.0	73.3	71.4	69.7	65.8	70.2	67.5	68.0	67.5	66.4	92.0
Finland	100.6	108.8	106.8	102.0	101.1	98.5	105.6	108.7	119.9	114.1	97.4	100.0
Sweden	102.0	107.1	100.7	101.2	96.6	94.5	95.4	96.8	97.9	96.4	92.6	104.0
United Kingdom	91.1	93.8	90.7	90.1	86.2	86.4	86.8	84.2	85.0	84.7	84.3	87.5
Croatia	70.4	72.3	77.1	78.0	81.3	81.1	84.7	88.6	93.3	94.0	95.5	95.0
Turkey	129.8	142.2	149.9	150.6	150.7	164.0	153.6	158.0	167.0	172.7	184.0	–
Iceland	93.6	96.1	101.0	105.2	111.5	109.9	109.5	109.9	107.9	109.7	110.5	110.0
Liechtenstein	86.9	86.8	86.8	86.8	86.8	86.8	86.8	86.8	105.3	–	–	92.0
Norway	100.2	106.0	105.9	106.4	108.4	107.6	110.2	107.6	109.0	110.3	108.8	101.0

– = Not available, a = Break in series

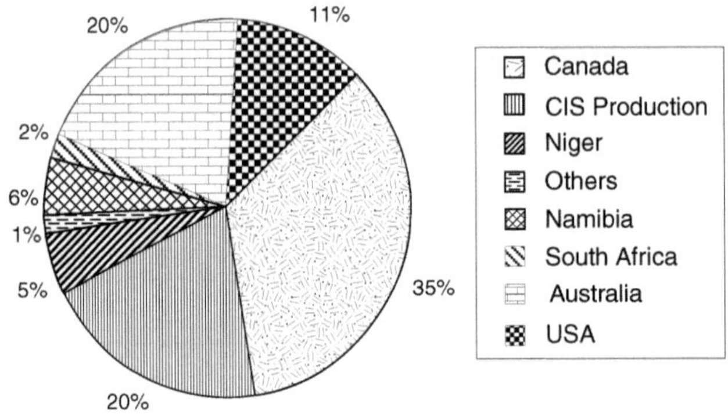

Fig. 1. *Cumulative world uranium production (source: IAEA).*

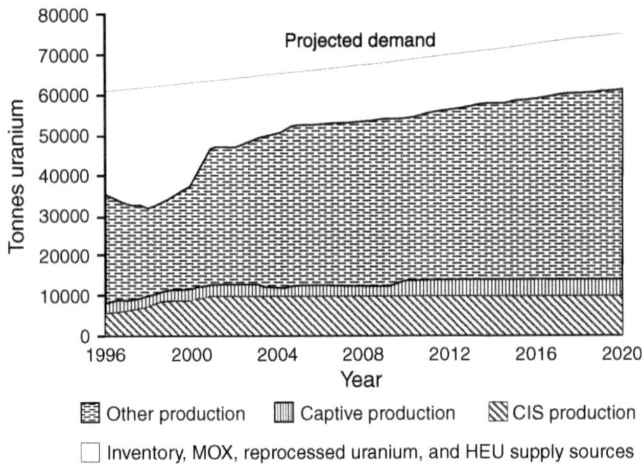

Fig. 2. *Uranium production and supply forecast, 1997–2020 (source: IAEA).*

5.2.3.3. Biofuels: pros and cons

Keeping in mind the high amount of CO_2 emitted from the energy sector, one cannot neglect the fact that biofuels are a cleaner and environmentally friendlier alternative to oil and natural gas. However, doubts over the environmental and social benefits of biofuels and their economic competitiveness prevail. Biofuels are subject to criticism on several points:[103]

First, land-use changes associated with increasing production may speed up the destruction of natural habitats. There are cases of deforestation in Indonesia and Malaysia for increasing the production of palm oil used for the production of biofuels. Regarding the security of supply, the figures (in particular for Europe) are still disturbing. In the case of the use of first-generation biofuels, about 50% of the supplies will have to be imported. Biofuel industries based on so-called "first generation technology" are raising agricultural commodity prices. The second generation of biofuels will probably become widely available around 2015. Another point of criticism is the direct impact on climate change, as in some cases, rainforests have been cut down for the purpose of using the land to grow energy crops for biofuels.

Furthermore, an increase in biofuel production may decrease the availability of freshwater for alternative uses, as the production process requires high quantities of water. A typical ethanol factory producing 50 million gallons of biofuel per year needs about 500 gallons of water a minute.[106]

The trade-off between an eventual decrease in CO_2 emissions and a wider use of biofuels is a zero sum game rather than a successful strategy to fight climate change, considering all that it entails such as higher food prices, deforestation and huge subsidies which could be otherwise transferred to less controversial renewable energy sources. The UN Food and Agricultural Organization (FAO) already warned that displacing food production has a disastrous effect on food security in developing countries.

The environmental and economic efficiency of biofuels is increasingly doubted, because their contribution to GHG emission reduction is only marginal while at the same time, world food production and thus the existence of people in developing countries is being threatened by the increasing use of biofuels.

5.2.3.4. Kyoto and Bali – missed chances

The decline in the availability of conventional energy sources and the responsibility to tackle climate change are intensifying the need for an international agreement and a stable legal basis for finding a common solution to the problems. However, all efforts to establish a binding legal basis that could ensure that the developed countries will meet their responsibilities have failed. The Kyoto protocol to the United Nations Framework Convention on Climate Change was adopted on 11 December 1997, but the USA has not ratified it yet. The EU adopted the world's first Emissions Trading System (ETS) in order to help attain its commitments under the Kyoto Protocol. ETS sets a cap on CO_2 emissions from energy-intensive industries and allows member states to buy and sell surplus allocations on an EU-wide market.[107]

The only result from the major climate conference in Bali in December 2007 was the 'roadmap' defined for further negotiations towards a successor to the Kyoto Protocol, which expires in 2012. The negotiations failed to establish a mechanism for a real burden-sharing in the fight against climate change, thus missing another chance for cooperation.

5.2.3.5. The demand to ensure energy supplies for third world countries

According to predictions of the World Energy Outlook, in the next 25 years, energy consumption will increase by 60%, with the major part of this growth being due to increased demand in developing countries.[108] The renewable energy share, although growing, is projected to stay rather unchanged with 14%, although there is a huge need for additional energy sources for the 1.4 billion people who will be without electricity in 2030.[109] In this regard, an essential step towards the goal of sustainable and clean energy supplies is to ensure a technology transfer to the developing countries for using the planet's available natural resources like the sun, water and heat for generating clean solar, wind and geothermal power.

5.2.4. Competitiveness – tackling climate change without affecting economic growth

Completely rejecting the use of fossil fuels is still rather unthinkable, not least because the adoption of a greener technology would imply a lot of additional costs for the industrial sector. An appropriate solution for the use of the world's coal reserves is the CCS technology. Investment in this technology could ensure a cleaner use of fossil fuels.

The CCS technology, referred to as carbon sequestration, presents a promising way for large-scale reductions in CO_2 emissions by capturing carbon dioxide (CO_2) from large point sources (such as fossil fuel power plants) and injecting it underground instead of releasing it into the atmosphere. The idea seems quite appropriate, as taking fossil fuels completely out of the energy mix seems rather impossible, at least in the short and medium term. Renewable energy sources still account for only a minor portion of the current energy mix and the situation does not seem to be changing drastically – at least in the short term – while world energy demand is continuously growing. The CCS technology is already commercially available, but because of the high costs, heavy public support will be required if it is to be put in practice widely. Some countries, including Norway, are convinced of the fundamental significance of this technology for mitigating climate change and are already subsidising the plants.[110] As reasonable places to store the CO_2, some researchers are pointing to the oil and gas fields, many of which are becoming emptied of hydrocarbons.

5.2.5. Conclusions

In light of the large number of urgent problems emerging at the crossroads between an era of over-reliance on fossil fuels and a time when evidences of climate change and the consequences of human actions are becoming more and more obvious, it is time to think of comprehensive solutions. Having listed the current multidimensional problems in the field of energy policy, one can conclude that the diversification of the world's energy mix is urgently needed. There is a strong call for alternative and innovative sources of energy generation which could ensure sustainable, carbon-free and secure supplies. All technologies and supporting instruments – including space applications – will be needed to meet this need.

As a basis for the further development of energy policy, the creation of a broad legislative framework is needed. This legislation should encompass not just a number of largely isolated sector-specific policies, but rather a comprehensive, outcome-oriented framework including the contributions of different sectors. Putting together a package of disconnected policies carries the risk of yielding less than optimal outcomes and ignores the interaction between policies.

In this regard, we should bear in mind technology-based, futuristic and even visionary solutions, however bold they may seem today. Space could become an alternative for a new source of energy generation. We do not have the time to question possible solutions and to be suspicious about their contribution. Space could provide new solutions for energy generation. However, if this is going to happen, large investments and a stable financial framework will be needed to support further research and future applications.

---

[91] International Energy Agency; Organisation for Economic Co-operation and Development. World Energy Outlook 2006. Paris: IEA; OECD, 2006.

[92] Hennicke, Peter and Müller, Michael. Weltmacht Energie: Herausforderung für Demokratie und Wohlstand. Stuttgart: Hirzel, 2005. p. 72.

[93] Umbach, Frank. "Sichere Energieversorgung auch in Zukunft: Die Notwendigkeit einer europäischen Strategie." Internationale Politik 59.8 (2004): 18.

[94] Javier Solana, EU High Representative for the Common Foreign and Security Policy, at the EU Energy Conference "Towards an EU External Energy" in Brussels, Belgium, 20 Nov. 2006.

[95] Commission of the European Communities; Secretary General; and High Representative for the Common Foreign and Security Policy. "An External Policy to Serve Europe's Energy Interests." Paper from Commission/SG/HR for the European Council. S160/06, 2006.
[96] Amineh, Mehdi P. Globalization, Geopolitics and Energy Security in Central Eurasia and the Caspian Region. The Hague: International Energy Programme, 2003.
[97] Widdershoven, Cyril. "A Gas OPEC in the Making?" European Energy Review 1.4 (2008): 36.
[98] British Petroleum. Statistical Review of World Energy. London: BP, 2001 and 2005.
[99] Flavin, Christopher. "Wind Power Sets New Record in 1998: Fastest Growing Energy Source." Worldwatch Institute Online 29 Dec. 1998. http://www.worldwatch.org/node/1641
[100] "Wind Power Growth Blows Past Projections." Worldwatch Institute Online 2008. 1 Dec. 2008. http://www.worldwatch.org/node/5698
[101] Commission of the European Communities. "An Energy Policy for Europe." Communication from the Commission to the European Council and the European Parliament. COM (2007) 1 final, 10 Jan. 2007.
[102] Commission of the European Communities; see endnote 11.
[103] Commission of the European Communities. "Limiting Global Climate Change to 2 Degrees Celsius: The Way Ahead for 2020 and beyond." Communication from the Commission to the Council, the European Parliament, the European Economic and Social Committee and the Committee of the Regions. COM(2007) 2 final, 10 Jan. 2007.
[104] A commitment to this target was first made in the Council Conclusions of 25 June 1996.
[105] Behrens, Arno. Biofuels or Bicycles? Why the European Union Should Reconsider Its Biofuels Target. CEPS Commentary, 20 Feb. 2008. 1 Dec. 2008. http://shop.ceps.eu/BookDetail.php?item_id=1617
[106] "Ethanol and Water: Don't Mix. New Reasons to Be Suspicious of Ethanol." The Economist, 28 Feb. 2008. 29 Feb. 2008. http://www.economist.com/world/unitedstates/displaystory.cfm?story_id10766882
[107] See: Emissions Trading Directive 2003/87/EC. In 2007, the Commission reviewed the EU ETS to ensure that emissions trading would reach its full potential; cf. Commission of the European Communities. "An Energy Policy for Europe." Communication from the Commission to the European Council and the European Parliament. COM (2007) 1 final, 10 Jan. 2007. p. 11.
[108] Expert Group on Renewable Energy. Increasing Global Renewable Energy Market Share: Recent Trends and Perspectives. New York: United Nations Department of Economic and Social Affairs, 2005. 1 Nov. 2008. http://www.un.org/esa/sustdev/sdissues/energy/op/beijing_re_egm/beijing_re_report.pdf
[109] Ibid., p. 11.
[110] The largest CO_2 storage project to date is located on the Sleipner field in the North Sea.

5.3 Remote sensing for sustainable hydrocarbon exploration and reservoir monitoring

Andreas Laake

5.3.1. Introduction

The decline of fossil fuel reserves has shifted the focus of hydrocarbon exploration and reservoir monitoring to increasingly remote areas. Often, these areas are environmentally sensitive and difficult to operate in. Therefore, in-depth information about the area targeted for a project is required at an early stage of planning to ensure a sustainable and safe execution.

Remote sensing data provide the required information, because the data acquired through satellites are densely sampled images and digital elevation models that contain spatial information about a variety of aspects of the Earth's surface. When combining spatial images with the elevation information from digital elevation models, the information from remote sensing data can be represented in three dimensions.

Hydrocarbon exploration and reservoir monitoring use primarily electromagnetic data in the optical and microwave range. To integrate remote sensing data with digital elevation models and geophysical data from other data sources, all data are geographically referenced and stored in a geographic information system (GIS). Thereby, the complete information is available simultaneously for joint interpretation.

Answers from space are available for all phases of hydrocarbon exploration and reservoir monitoring projects, and range from environmentally sustainable design, geological modelling, logistic risk estimation and geophysical data quality estimation, to reservoir monitoring.

5.3.2. Hydrocarbon exploration

Finding hydrocarbons and producing reservoirs in a sustainable fashion is of major importance for securing the future energy supply. The decline of hydrocarbon reserves is leading to operations in increasingly complex and remote areas. Minimising the environmental impact of exploration and production have become essential parameters in the responsible operation of the hydrocarbon industry.

The mapping and monitoring of hydrocarbon reservoirs requires the three-dimensional imaging of subsurface structures over time. Acoustic methods such as surface seismic surveying are the only techniques that provide the required complete, three-dimensional subsurface spatial data. Borehole and logging methods provide complementary data for reservoir monitoring. Watson et al.,[111] Barnes,[112] and Mathieu[113] described applications of remote sensing technology to assist in the exploration for minerals and hydrocarbons.

5.3.3. Challenges for risk minimisation and sustainability

Reservoir mapping strongly depends on the quality and accuracy of the geophysical image and its changes over time. The sources and receivers employed in surface geophysical methods are affected by the character and properties of the Earth's surface. Variations in elevation and elastic properties may introduce errors in seismic data that result in degraded information about the reservoir. The key challenge for reservoir mapping is to compensate for the distorting effects which the surface and the near surface have on the data quality. Their correction is therefore crucial for the success of the surface geophysical techniques.

Reservoir monitoring is essential for the optimum drainage of a reservoir. The critical parameter is the fluid flow within the reservoir. Mapping the fluid flow requires even more accurate data because small changes in the reservoir geometry as a result of production must be detected. Surface seismic, borehole, and well-logging data alone are often not sufficient to achieve this task.

The environmental sustainability of exploration and production are of significant importance because hydrocarbon reservoirs are increasingly located in remote and sensitive environments. Understanding environmental threats allows us to design the survey and the surface facilities, while minimising the environmental impact.

The main challenge is to minimise exploration and production risks. Remote sensing data can provide high-resolution spatial information that complements geophysical data. The application of remote sensing to seismic exploration has been described by Insley and Laake.[114,115] Geological structures[116,117] and near-surface properties[118] have been mapped successfully from satellite data. A sustainable design of the geophysical surveys based on remote sensing data has been proposed by Xinguo et al., and Laake and Insley.[119,120] Interferometric spaceborne radar techniques have been applied to the mapping of small surface elevation changes above hydrocarbon reservoirs.[121,122]

5.3.4. Data provided by remote sensing technology

Remote Earth observation employs a variety of electromagnetic waves to measure absorption or scattering for mapping and characterising the surface.[123] Table 4 gives an overview of the data and their applications.

A digital elevation model (DEM) may be obtained from radar or Lidar data. It provides a three-dimensional map of the surface. The DEM gradient delivers the terrain slope, which highlights steep escarpments.

Optical data are acquired through passive sensors measuring the reflection and absorption properties of the surface. Visible optical data are obtained from high-resolution satellite sensors. Their main application is the mapping of the details of the surface cover and infrastructure, using natural colours. Multispectral optical data are provided by satellite sensors with different spectral bands comprising mostly visible, near, and shortwave infrared as well as thermal infrared data. Surface materials exhibit different absorption characteristics in different bands. Combinations of bands to continuous colour (RGB) images reveal features such as surface rocks and land use.

Tab. 4: *Remote sensing data and their applications.*

Data	Wavelength	Resolution	Applications
Digital elevation model	n/a	1–100 m	Topography, terrain gradient
Visible optical data	400–700 nm	<1–30 m	Infrastructure and surface feature mapping
Multispectral optical data	400 nm to 100 µm	15–90 m	Land use, surface characterisation
Microwave data	1–10 cm	1–30 m	Surface roughness, moisture mapping
Interferometric synthetic aperture radar data	1–10 cm	1–30 m <1 cm vertical	Surface movement

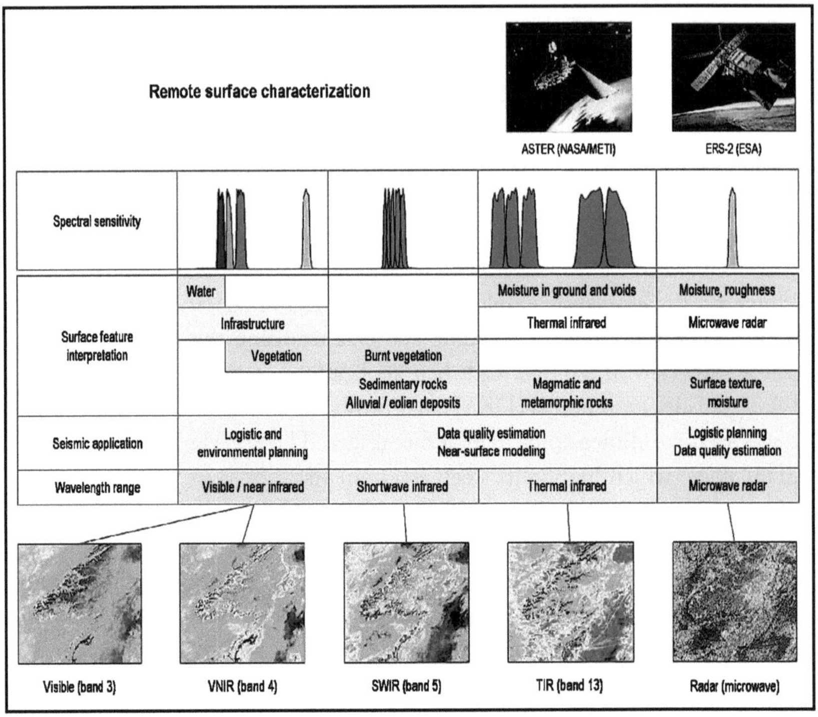

Fig. 3. *Surface characterisation from different remote sensing data.*

Microwaves are emitted from an active satellite. They are back-scattered either at the surface or in the layers directly beneath the surface. Microwave data reveal the roughness of the surface texture and provide information about moisture in the near surface. Synthetic aperture radar (SAR) data may be used in an interferometric approach to measure small surface movements of less than 1 cm over large areas. Figure 3 summarises the different types of remote sensing data and their applications.

5.3.5. Surface and near-surface mapping

The interpretation of remote sensing data regarding near-surface physical properties is based on land use, lithology, and roughness mapping.[124] In desert terrain, this interpretation is mostly straightforward because rock formations or soils are exposed at the surface and are readily accessible for remote sensing. In the case of partial coverage with vegetation, the interpretation of the satellite data for surface

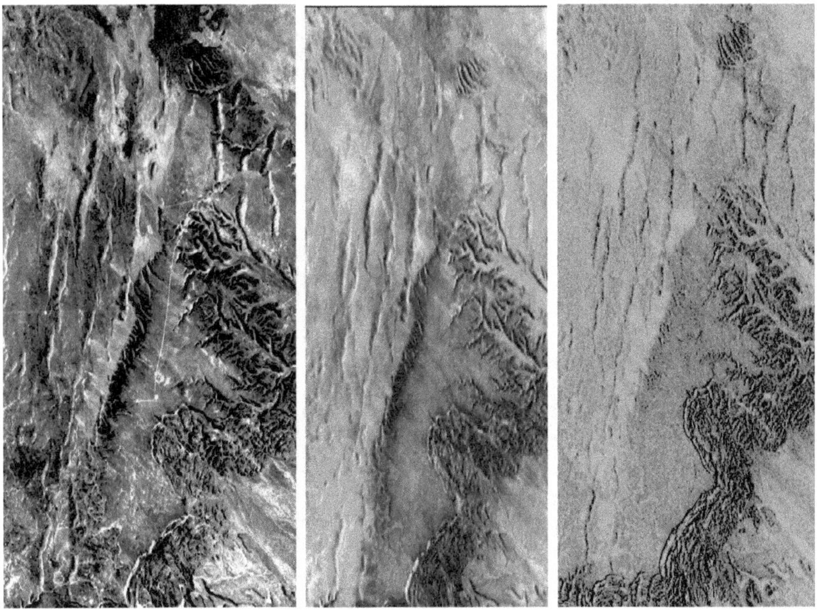

Fig. 4. *Preprocessing of multispectral data to remove the vegetation effect. Raw RGB image (left), RGB image comprising thermal infrared data (centre) and roughness from radar data (right).*

lithology is altered by the spectral response of the vegetation, because individual plants are not resolved. In this case, the data must be pre-processed to remove the distortion caused by the vegetation. Figure 4 gives an example from a basalt plateau with sparse vegetation between basalt blocks. The raw RGB image shows a uniform blue-green tone for both a block basalt area with sparse bush on the right, and fine gravel with grass on the left (left image). Using the different responses of vegetation to visible and infrared data, the surface covered with basalt blocks can be distinguished from the fine basalt gravel (centre image). The result is confirmed by the higher roughness signature (orange-red area in the right image).

5.3.6. Data integration

Finally, the interpreted satellite imagery is merged with geological models. In the case of reservoir monitoring, the surface movement is correlated with the subsurface geological structure to obtain information about the fluid movement in the reservoir.[125] For exploration, the lithology interpretation is correlated with

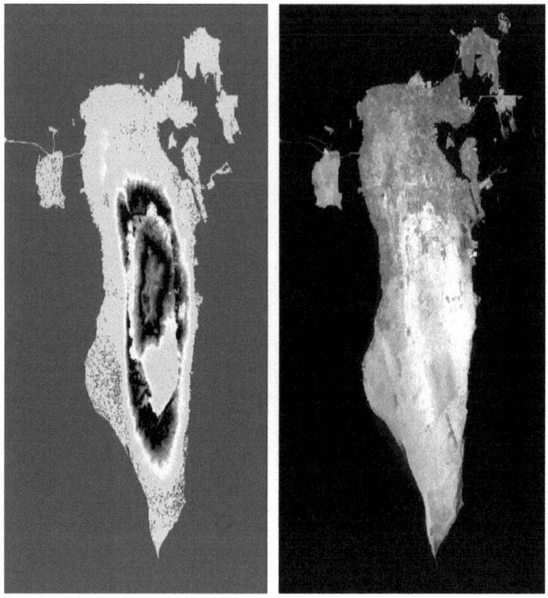

Fig. 5. *Integration of remote sensing data with DEM and geological data in a GIS database.*

geophysical measurements to infer elastic parameters required for geophysical data quality estimation.

This integration is preferably done in a geographic information system. All data layers are projected to the same grid and co-referenced, which allows the joint interpretation and correlation of the individual map layers. In combination with the DEM, this map set forms a 3D-model of the surface and near surface (Figure 5). The GIS database also serves as an archive for the various map data.

5.3.7. Products

Remote sensing can provide a variety of products for hydrocarbon exploration and reservoir monitoring. Surface and near-surface characterisation may assist in exploration by mapping logistic risks. The map sets archived in the GIS database also provide terrain information to help plan the survey and, thereby, mitigate the risk. The information about surface features and land use helps in assessing the environmental impact and assists in optimising the logistics so as to minimise the environmental impact. The estimation of the source- and receiver data quality is another important answer that remote sensing data can provide.[126,127]

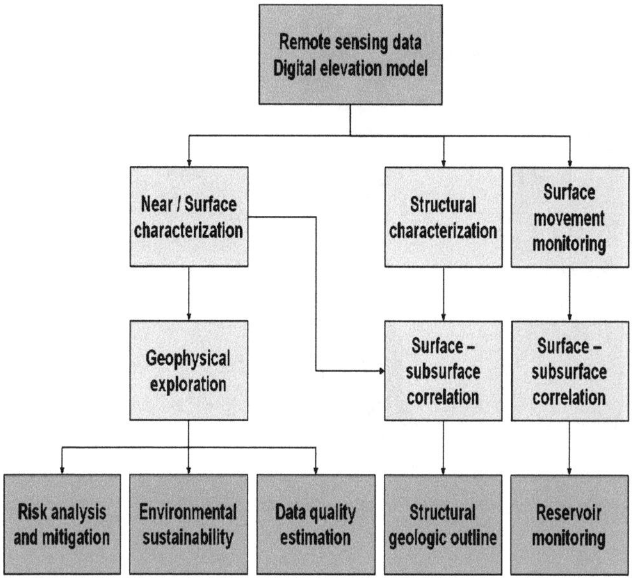

Fig. 6. *Workflow for remote sensing-based products for hydrocarbon exploration and reservoir monitoring.*

The correlation of the surface and near-surface results from remote sensing with structural interpretation allows for the interpretation of the outcrop geology to a depth below the surface.[128,129] Surface movement monitoring from interferometric SAR data may be correlated with 3D-geophysical measurements and the subsurface geology to attribute surface movements to reservoir fluids movements. This technique is particularly successful for shallow reservoirs. Figure 6 summarises the products derived from remote sensing data.

The special processing of multispectral optical data may reveal multiple aspects of the surface characterisation. Figure 7 shows two examples of the interpretation of remote sensing data.

The left image shows a simple continuous-colour RGB image in a band combination that highlights different forms of land use, such as green for vegetation, grey to purple for infrastructure, light yellow for limestone, and light red for clastic sediments. An area of coastal salt plain (sabkha) is highlighted by the colour cyan.

The right image shows a set of band combinations that reveals a more detailed composition of the surface. This image was processed with a focus on the geological structure (red-orange) which is partly hidden under the infrastructure in the northern part of the island (grey-green). Also, a karst anomaly is highlighted as a purple square in the south of the island. Sabkha (yellow), infrastructure (grey-green), and farmland (dark-green) are of particular importance for logistic

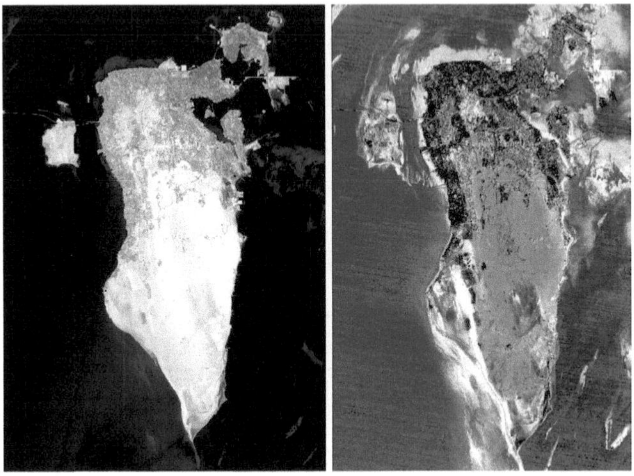

Fig. 7. *Examples of terrain-interpreted remote sensing data: Land use (left) and surface characterisation (right).*

planning as well as for the data quality. Sabkha and farmland are sensitive terrain in a desert environment that requires special attention for minimum impact. Mapping and characterising such areas prior to any exploration project ensures

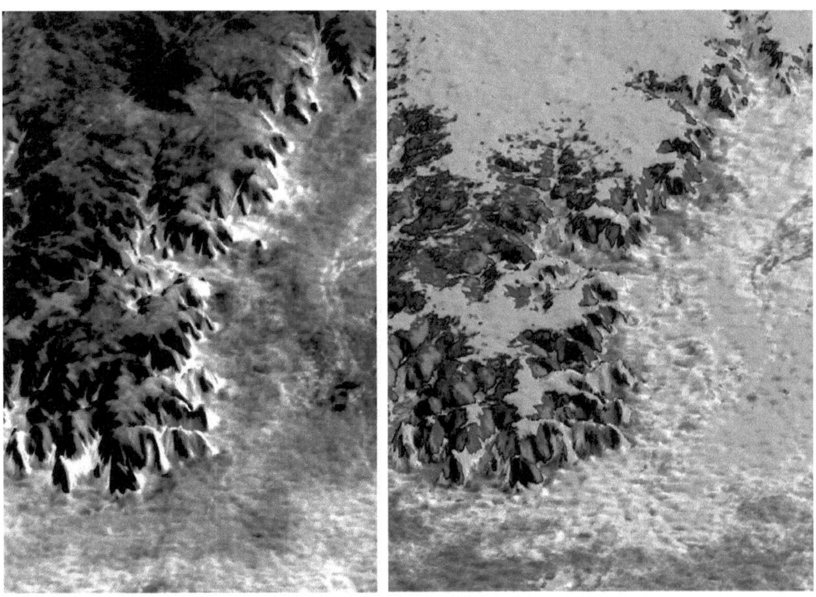

Fig. 8. *Example of terrain interpretation for lithology (left) and risks for logistics, environmental impact, and data quality (right).*

that the logistic effort for geophysical operations can be minimised in such areas and be compensated by efforts in other areas of the survey, while also maintaining the data quality.

The second case study shows an interpretation of surface rock types and elastic properties (Figure 8). A steep limestone plateau (red areas) can easily be distinguished from blue-green sand and clastic rocks. Furthermore, aeolian deposits of gypsum and fine sand are mapped at the bottom of the limestone plateau (left image).

The right image shows the conversion of the lithologic characteristics into the elastic properties important for geophysical operations. Blue and blue-green represent areas of hard limestone, which is characterised by high elastic velocities and a poor coupling of the geophysical sources to the ground. Red and yellow areas mark soft clastic terrain, which shows relatively slow elastic velocities and a high attenuation for elastic waves. The more intense the red, the softer the surface, which is often associated with wadis. Purple to pink areas delineate steep and thus dangerous terrain that should be avoided by vehicles because of the risk of vehicle rollover.

5.3.8. Conclusion: answers by remote sensing technology

Remote sensing data provide continuous map information about a variety of parameters critical for the efficient identification, sustainable exploration, and optimised recovery of hydrocarbon resources. Electromagnetic remote sensing data are used that comprise visible and multispectral optical data from passive

Tab. 5: *Remote sensing-based answers for the hydrocarbon industry.*

Answer	Remote sensing data	Method
Prospect identification	Multispectral data	3D geological interpretation
	Geophysical surveys	
Logistic planning	Multispectral/radar data	3D terrain characterisation
Data quality estimation	Multispectral/radar data	3D terrain characterisation
Environmental sustainability	Multispectral data	Surface characterisation
	Reservoir management	
Infrastructure planning	Multispectral data	3D terrain characterisation
Reservoir monitoring	Radar data	Interferometric SAR

satellites and microwave data from radar satellites. In cases where 3D-information is required, the remote sensing data are combined with digital elevation models. The technologies applied for extracting the information employ image processing, spectral analysis, and interferometry. Often, the modelling supports the interpretation, particularly in 3D. Table 5 summarises the answers provided by the specific type of data and the methods used to extract the desired information.

Critical to the success of remote sensing-based answers for the hydrocarbon industry is the geographic referencing of all data sets and their integration in a geographic database such as GIS. The representation of multiple characteristics of the surface and the subsurface as map layers enables the user to interpret the data for surface and subsurface geology, logistic risks, geophysical data quality, and surface movement. The environmental impact of the technology preferred for a given application can be assessed prior to planning the project, which allows the adaptation of the survey for achieving environmental as well as economic targets while delivering an adequate data quality.

Remote sensing data provide unique spatial map information essential for the efficient planning and monitoring of hydrocarbon exploration and reservoir monitoring projects. The demand for environmentally sustainable operation in increasingly remote and sensitive terrain can be addressed by designing and monitoring the projects in a spatial GIS database. Ultimately, remote sensing provides the means for optimising hydrocarbon production with respect to environmental sustainability and economic viability.

Acknowledgements

The author thanks WesternGeco for permission to publish this article. The European Space Agency supported this work under the Earth Observation Market Development Programme.

[111] Watson, Kenneth, Kruse, Fred A., and Hummer-Miller, Susanne. "Thermal Infrared Exploration in the Carlin Trend, Northern Nevada." Geophysics 55.1 (1990): 7–79.
[112] Barnes, Michael. "High Resolution Satellite Imagery Applied to Oil and Gas Projects." Expanded Abstracts, 75th SEG Annual International Meeting, 2005. 2345–2347.
[113] Mathieu, Pierre-Philippe. "Satellite Data Finding More Uses in E&D, Other Industry Segments." Oil and Gas Journal 104.3 (2006): 44–48.
[114] Insley, Martin and Andreas Laake. "Satellite Based Seismic Technology, Case Study: Berkine Basin, Algeria." Extended Abstracts, 66th EAGE Conference & Exhibition, 2004. Paper E032.
[115] Laake, Andreas and Insley, Martin. "Using Satellite Imagery Aids Seismic Surveys in Planning, Acquisition, and Processing." World Oil 225.9 (2004): 27–33.

[116] Fraser, Andrew J., Huggins, Paul, Cleverley, Paul H., and Rees, John L. "A Satellite Remote-Sensing Technique for Geological Horizon Structure Mapping." SEG Technical Program Expanded Abstracts, 65th SEG Annual International Meeting, 1995. 134–137.
[117] Offield, Terry W., Abbott, Elsa A., Gillespie, Alan R., and Loquercio, Sabino O. "Structure Mapping on Enhanced Landsat Images of Southern Brasil: Tectonic Control of Mineralization and Speculations on Metallogeny." Geophysics 42.3 (1977): 482–500.
[118] Laake, Andreas and Cutts, Andrew. "The Role of Remote Sensing Data in Near-Surface Seismic Characterization." First Break 25.12 (2007): 61–65.
[119] Xianguo, Huang, Xueqiang, Chen, Shujie, An, and Zhaolong, Xu. "3D Seismic Acquisition in Complex Mountainous Area, YKB, Western China." SEG Technical Program Expanded Abstracts, 74th SEG Annual International Meeting, 2004. 87–90.
[120] Laake, Andreas, and Insley, Martin. "Applications of Satellite Imagery to Seismic Survey Design." The Leading Edge 23.10 (2004): 1062–1064.
[121] Gabriel, Andrew K., Goldstein, Richard M., and Zebker, Howard A. "Mapping Small Elevation Changes over Large Areas: Differential Radar Interferometry." Journal of Geophysical Research 94.B7 (1989): 9183–9191.
[122] ud din, Saif, Al-Dusari, Ahmad, Al-Ghadban, Abdulnabi, and Aritoshi, Mio. "Use of Interferometric Techniques for Detecting Subsidence in the Oil Fields of Kuwait Using Synthetic Aperture Radar Data." Journal of Petroleum Science and Engineering 50.1 (2006): 1–10.
[123] Sabins, Floyd F. Remote Sensing: Principles and Interpretation. 3rd ed. New York: W.H. Freeman and Co., 1997.
[124] Laake, Andreas and Cutts, Andrew see endnote 28.
[125] Xu, Haibin and Nur, Amos. "Integrating Reservoir Engineering and Satellite Remote Sensing for (True). Continuous Time-Lapse Reservoir Monitoring." The Leading Edge 20.10 (2001): 1176–1180.
[126] Insley, Martin and Laake, Andreas. "Seismic Quality Analysis in Algeria: Application of Earth Observation Data Sets to Oil & Gas Exploration." PESGB London Evening Meeting, 2003.
[127] Laake, Andreas and Tewkesbury, Andrew. "Vibroseis Data Quality Estimation from Remote Sensing Data." Expanded Abstracts, 67th EAGE Conference & Exhibition, 2005. Paper G017.
[128] Offield, Terry W., et al.; see endnote 27.
[129] Sabins, Floyd F.; see endnote 123.

5.4 Sustainable energy supply on Earth – can space contribute?

Wolfgang Seboldt

5.4.1. Introduction

A key challenge for mankind in the coming decades will be the sufficient supply with energy and the corresponding employment of environmentally friendly technologies. The average present primary world power consumption is about 15 TW (TW = Terawatt = 10^{12} W), and demand will probably increase to 22 TW by 2030 due to the growing population and further industrialisation.[130–133] The electric share amounts to ~2.0 TW today, with a projected share of 3.9 TW in 2030 (see Table 11, which also displays the European power consumption).

Presently, about 85% of the primary world power is fossil-fuelled, with significant CO_2 emissions into the atmosphere. There is a strong requirement for new solutions and a change in energy policy due to the following reasons:

- The share of CO_2 in the atmosphere is growing, and the future deposition must at least be stabilised within the next decades and possibly reduced in order to avoid dramatic consequences from the greenhouse effect.
- Fossil (oil, gas, coal) and conventional nuclear fuels (for fission) will be depleted in the long run (around 2100 or not far beyond); in addition, the safe storage of waste from nuclear power plants is an unsolved problem.
- The growing power demand – especially in developing countries – must be met somehow.

Regenerative (also called renewable) forms of energy – i.e. solar/wind/hydro power and to some extent energy from biomass as well as from geothermal

Tab. 6: *World/European power demand (source: IEA).*

TW[= 10^{12} W]	Primary power [TW]	Electric power [TW]
In 2004	15/2.3	2.0/0.4
In 2030	22/2.6	3.9/0.5

sources – are generally considered important sustainable terrestrial alternatives for the future. Also, controlled fusion may play a significant role in the long run if research is successful and the technology safe, but operational power plants cannot be expected before 2050 at the earliest. Obviously, regenerative solutions must be considered with higher priority for the coming decades.

This paper will try to address the following question: can space contribute to a sustainable global energy supply on Earth? Solar power – collected in space and transmitted to the Earth's surface – has been proposed for many years as a potential space option. Also extensively discussed in the literature is ^{3}He-mining on the Moon, to be used for future cleaner terrestrial fusion power plants. Both space options – often claimed as arguments in support of enhanced future space activities – will be analysed in terms of their technical and economic feasibility in Section 4.3, also considering the involved time scales for eventual realisation. But first, we will shortly review the status and perspectives of the terrestrial regenerative power supply in Section 4.2, because potential space options must at least compete with these upcoming terrestrial solutions.

5.4.2. Terrestrial solutions for a global regenerative power supply

The solar radiation flux in space at Earth's distance is ~1368 W/m^2. It integrates to more than 10^5 TW of power inflowing onto the whole Earth (including its atmosphere), about a factor of 10^4 more than the primary world power demand. After interaction with the atmosphere and surface, nearly all of it is reradiated into space at modified wavelengths, maintaining a thermal quasi-equilibrium on Earth. A fixed (co-rotating) area on the surface of the Earth receives a typical continuous solar radiation flux between 100 and 300 W/m^2, averaged over day/night and year, and depending on the region (e.g., the larger values are obtained closer to the equator).[133,134] The transformation into useful forms of power (e.g. electricity) is typically performed via photovoltaic (PV) or solar thermal (ST) devices.

The past development of PV cells from single junction Si to multiple junction GaAs cells resulted in an increase in the conversion efficiency to electricity from 10% to 30% or even more. Efficiencies from 5% to 15% are achieved by thin-film solar cells (TFSC) made of amorphous silicon or other materials like CdTe and $CuInSe_2$. They promise low production and power costs for the future due to low material needs, simple production processes and reduced requirements for frames and trusses. The yearly solar cell production worldwide went up from 300 MW_p

(peak electric power) in 2000 to more than 2.5 GW_p in 2006, a dramatic increase but still insignificant in view of the world's power demand.[135]

In ST plants, the incoming solar radiation is concentrated and converted into heat, which is then used in conventional turbo-machinery to produce electricity. There are two main concepts: parabolic troughs and central receivers (see Figures 9 and 10). ST plants with troughs are being run up to hundreds of MW_p (e.g. SEGS in California, USA, with an area of 2.5 km^2). Central receiver (or solar tower) power plants have reached more than 20 years of experimental operation time (e.g. Barstow, California, USA, or Almería, Spain). Different heliostat and receiver/

Fig. 9. *Parabolic troughs of the Solar Energy Generating Systems (SEGS) in California's Mojave desert (source: Solel Solar Systems).*

Fig. 10. *Central receiver (solar tower) in Almería/Spain (source: DLR).*

absorber techniques including heat transfer media have been tested with significant improvements for the operation strategy, safety and energy storage concepts (see below). Also, combined cycles and (solar/fossil) hybrid systems are under investigation, aiming at better system efficiencies (>15%) and mid/full-load operation.

Since in terrestrial solar power plants, conversion efficiencies to electricity of 15–30% are achieved or can be expected in the near to medium-term future, average power densities of 15–90 W_e/m^2 for electricity generation on Earth are attainable.[132] Even considering only the desert areas (~0.3% of the Earth's surface), installations of PV or ST systems on a few percent of the deserts would be sufficient for satisfying the world's electricity demand.[131,134]

The incoming solar radiation is also partially transformed into wind power (~10^3 TW) and – within the water circulation – into hydro power (~6 TW).[132,133,136] Geothermal heat (practically inexhaustible but not regenerative) is difficult to exploit, and biomass has a low power density of ~0.6 W/m^2.[137] Nevertheless, the estimated global integrated geothermal heat-flow is several 10 TW, and power from biomass is theoretically as large as 20 TW worldwide if parts of the farm lands (< 20%) are used together with major parts of up-to-now unused land.[132] While the numbers illustrate the considerable potential of these energy resources, the present input into the electricity grid is still relatively small (e.g. contributions to German electricity generation in 2007: ~6.5% from wind power, ~3.5% from hydro power and 3.8% from biomass, leading to a total contribution from regenerative sources to the German primary energy consumption of 6.7%).[138] The typical maximum output from a large wind turbine is in the order of MW_p. In favourable regions – working at high loads (30–40%) – electric power densities from wind power plants (~15 W_e/m^2) may not be far off from those of terrestrial solar power plants.[132] However, satisfying global electricity demands with wind power alone would require, however, a huge number of turbines on land and offshore with large areas for installations, taking into account the intermittent availability of wind power. But a significant contribution to a global energy-mix could certainly be expected. Besides solar/wind/hydro power, the mix should also incorporate biomass and geothermal sources, although probably only to a limited degree. Obviously, for reasons of environmental protection, only fractions of the abovementioned global regenerative potentials can actually be used, but all of them can contribute significantly to the required power supply. An adequate mix would be capable of satisfying – at least in principle – all present and future global power requirements.[139]

In the past, many of the regenerative options on ground were seen as unsuited for base load due to their discontinuous supply caused by day/night and seasonal cycles as well as unpredictable local weather conditions. With technological progress, however, quite obvious solutions became feasible: energy storage and the transmission of power over large distances. The latter enables the use of favourable sites for power production – often far from the end users – and also the 'smoothing' of the power output of renewable sources by connecting widely distributed power plants. The transmission of electricity via cables over large distances with small losses is a major issue (for the alternative of power relay satellites, see next section). Basically, this can be done by high-voltage (HV) alternating current (AC) or direct current (DC). HVDC transmission has reached a share of only a few percent world-wide by now.[140] It promises, however, lower transmission losses than HVAC lines (2–3% vs. 4–5% per 1,000 km) and also lower costs for distances of more than 800 km (overhead lines). Transmission voltage is expected to rise from currently 600–800 kV by 2020. Maximum transmission power for a single line or cable is 5–6 GW_e.

Energy storage with hydrogen – produced via the electrolysis of water – has been investigated in demonstration plants in the past (Euro-Québec Hydro-Hydrogen Pilot Project – EQHHPP; World Energy Network Using Hydrogen – WENET). Hydrogen production using direct concentrated solar radiation is under investigation at Almería/Spain (international project HYDROSOL II). Gaseous storage in large underground caverns is already in the commercial stage today and seen as being very promising in terms of safety and costs.[141] Gaseous H_2 pipelines of a length of several thousand kilometres for transmission are in operation. The storage of thermal energy is another option for which ST plants are ideally suited, with the corresponding technology – molten salt in tanks – being under development. Therefore, the continuous day/night operation of large ST systems is close to becoming operational, but due to cost considerations, combinations with pumped water storage – where possible from topography, etc. – appear promising, too. As already mentioned, hydroelectric power (from water reservoirs, dams, etc.) is an established component in the worldwide energy supply, and some of the corresponding facilities can be used for energy storage as well. A long-term alternative or complement to energy storage could be a Global Energy Network (GLEN).[131,133,134,142] In this concept, large ST plants located at favourable sites around the world would be connected to the major power consumers of the world via HVDC transmission lines and cables (total length: 82,000 km). The ST plants are dimensioned to cover the required electricity demand continuously at any time of the year. Thus, a kWh_e consumed in Europe could be produced by any of the ST plants worldwide and transmitted without intermediate power storage.

In summary, the technical challenges of efficient energy conversion (to electricity, hydrogen, heat, etc.) and the corresponding storage can be mastered, and all the abovementioned regenerative options have the potential to satisfy a significant portion of the global power demand. However, their large-scale use is still lacking. On the one hand, this is because of the present superiority of conventional power plants (oil, gas, nuclear) with regard to power generation costs, grid connectivity, power availability/flexibility, land use, independence from weather, etc. On the other hand, bulk regenerative power supply requires large areas of land and – in some cases – huge power storage facilities for night-time supply. This is a difficult undertaking especially for the industrialised areas in the temperate zones of the Earth. Access to suitable locations close to the equator or sites with favourable wind conditions needs to be guaranteed by political agreements. The necessary areas are certainly large, e.g. about 400 km × 400 km for solar power to satisfy the total world power demand, but this represents no fundamental obstacle. Because of the dramatic progress achieved in the past decade, electricity production via ST plants in base load mode (i.e. with night time storage) and via wind power plants has become nearly competitive with conventional (non-regenerative) power production and is expected to lie in the range of 0.05–0.15 €/kWh$_e$, promising even further reductions. PV systems may be more expensive by a factor of 3.[135,139] A recent ESA-study, aiming at the comparison of future terrestrial and space-based solar power plant scenarios for the timeframe 2020/30, concluded that ST tower plants with local hydrogen storage for continuous power supply promise costs of less than 0.1 €/kW$_e$h (scenarios with 500 MW$_e$ to 500 GW$_e$ plants).[143,144] It was also shown that the energy amortisation time (EAT) of PV or ST systems is around or less than 2 years.

5.4.3. Space options for a global terrestrial power supply

5.4.3.1 Space Solar Power (SSP)

In general, SSP for terrestrial use comprises the idea of collecting solar power in space, converting it to electricity and transmitting it with large high-power antennas – directly or via orbital redirectors – to Earth-based receivers or rectennas (= antennas + electric rectifiers). The following "classical" SSP concepts are under discussion: Solar Power Satellites (SPS), Power Relay Satellites (PRS) and Lunar Power Stations (LPS).

The concept of SPS was invented in 1968 by Peter Glaser.[145] He proposed to place large photovoltaic systems in space – especially in geostationary orbit (GEO) at a height of 36,000 km above the Earth's surface – and to transmit the gained electric energy to the surface using Wireless Power Transmission (WPT). Since then, numerous studies, papers and lectures have been worked out worldwide on this idea.[131,133,146–148] In order to supply power in the order of GW_e on the Earth (typical output of a terrestrial power plant), very large structures of tenths of square kilometres covered with PV cells have to be assembled in space to collect and transform the incoming solar radiation into DC electricity (solar collectors/ concentrators with turbo-machinery might be an alternative). The structures must be extremely light-weight and at the same time sufficiently stable to allow attitude control. The transmission to the ground with microwaves (considered frequencies: 2.45 or 5.8 GHz) would be performed via an antenna, but lasers are also under discussion. For conversion, a huge number of active components, e.g. magnetrons, each producing a few kW, together with carbon-fibre slotted wave guides or patch/micro-strip antenna elements may be used. The conversion efficiencies obtained with magnetrons today are up to 70%.[149] They are mounted on the main antenna structure which has typically the size of 1 km, corresponding to a beam width of ~10 km at the receiving site on the ground. For beam focusing and steering, the antenna elements must form a coupled phased array. The design and synchronisation of this phased array should be such that power losses in the emitting elements and side lobes from the antenna are minimised. This is a complicated problem (especially for high power) and to some extent not yet solved or experimentally verified. Moreover, thermal control and power management is a very critical aspect due to the mentioned conversion efficiencies and potential additional losses in the cabling.[150] Superconductivity may be one part of a possible solution (with high expense) but also efficient radiators could be necessary. On the ground, the microwave radiation is transformed back into useful DC electricity via many dipole antennas distributed over an area of ~100 km^2 (rectenna site) with efficiencies of about 80%, and fed into the power grid.[131]

One motivation for pursuing SPS is that the Sun's irradiation can be utilised more efficiently in space than on Earth due to the independence from the 24-h day-and-night cycle and weather conditions. In GEO, there is a permanent illumination of an SPS all year long – except for short periods during vernal and autumnal equinoxes. This implies a factor of almost 5 in favour of an orbital SPS (1368 W/m^2 average illumination) compared to the best ground solar power plants (300 W/m^2 average at best). But there is also a drawback: due to physical and safety considerations concerning the interaction of the microwave beam with the ionosphere as well as with life on ground, only a restricted flux can be allowed for the microwave power. If the present U.S. standard for continuous microwave exposure (~20 mW/cm^2, corresponding to 200 W/m^2) would be imposed as the upper limit on ground, the average allowed flux on the rectenna would be as small as ~100 W/m^2 (Figures 11 and 12). Thus, the incoming continuous microwave power density from SPS would be only half the average of the continuous natural solar power density on ground, leading at most to a doubling of the electricity

Fig. 11. *Artist's view of a ground receiving rectenna (source: PowerSat).*

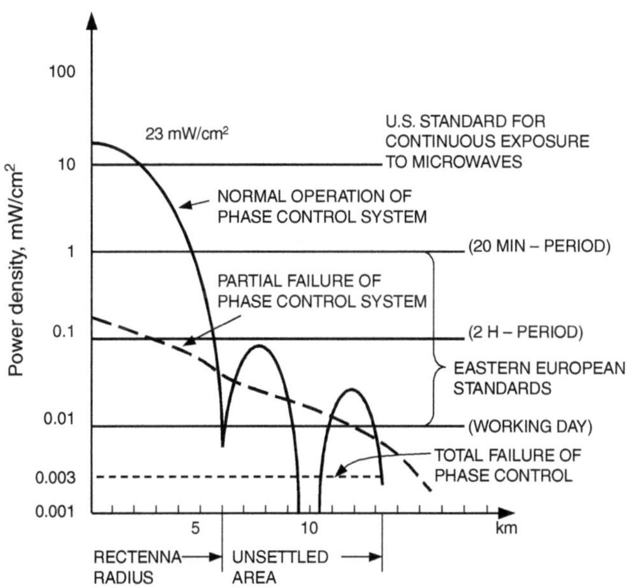

Fig. 12. *International standards for continuous microwave radiation exposure (source: NASA).*[131,158]

power output per square metre because of the higher efficiency of microwaves relative to electricity conversion. But the rectenna field must have an additional unsettled area for safety which may eliminate even this small advantage. On the other hand, we must keep in mind that terrestrial solar power systems are probably more complex and costly than dipole antennas for microwaves. In recent years, transmission via laser has also become a topic to be studied further. Of course, cloudiness would cause difficulties, and suitable safety standards would have to be defined and implemented, too.

The SPS concept was under thorough review and analysis by several U.S. companies and governmental organisations in the 1970s and 1980s.[151,152] The derived NASA "Reference System" formed the baseline for all following SPS concepts and studies. The power output of one satellite (10 km × 5 km × 0.5 km in size, 50,000 tonnes of mass) was designed to be 8 GW_e, the ground rectenna (~10 km) output was 5 GW_e. The satellites would be constructed by thousands of space-working astronauts and placed in GEO by a fleet of reusable, heavy lift launch vehicles (HLLV). These concepts – based on the conventional technology of those days – were deemed not very promising and the activity was put on hold. Since the 1990s, a renewed interest in concepts and technologies relevant to SPS has arisen due to significant progress in space technology (e.g. lightweight structures, PV, etc.).[153,154] In Japan, the SPS 2000 project was a larger attempt towards establishing an operational MW-class SPS in LEO.[147] The 1995–1997 NASA "Fresh Look" study generated a number of innovative architectural concepts and technologies aimed at reducing the required upfront investment.[146,145] In its 'Sun Tower' concept, NASA used inflatable, gossamer-like structures for the realisation of concentrators which direct the solar radiation onto PV receivers (Figure 13). The 'Sun Tower' with dimensions of 15 km × 0.2 km × 0.1 km and a total mass of about one 1,000 tonnes would be in sun-synchronous low-Earth orbit. About 100 inflatable concentrators would be mounted on a central truss structure, and a

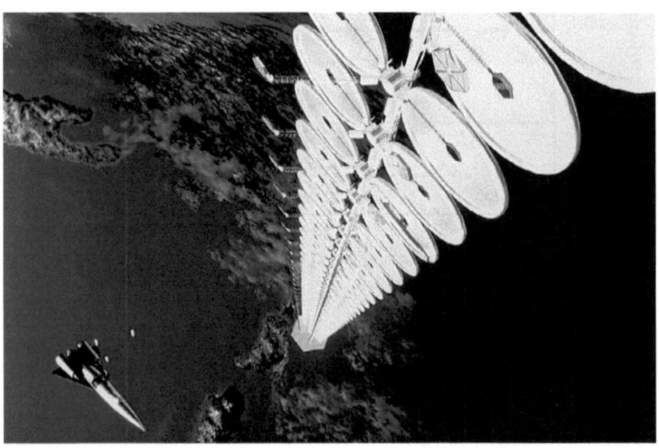

Fig. 13. *NASA 'Sun Tower' concept (source: NASA; artwork by Pat Rawlings/SAIC).*[146]

5.8 GHz antenna would transmit the power to the surface (ground output several hundred MW_e). The results of the "Fresh Look" study contributed to the decision of the U.S. government to provide NASA with 22 million dollars for space solar power research in fiscal years 1999 and 2000.[156] This was called the Space Solar Power Exploratory Research and Technology (SERT) programme led by the NASA Marshall Space Flight Center. Its continuation in 2001–2002 under the title of Space Solar Power Concepts & Technology Maturation (SCTM) programme[150] aimed at advancing key concepts and technologies for future SSP applications in the areas of WPT, intelligent robotics, power management/distribution and control, cost optimisation, and potential technology flight experiments.

In 2007, the U.S. National Security Space Office resumed the topic in a study and concluded that Space-Based Solar Power (SBSP) "presents an opportunity for strategic security (. . .) with the potential to be a disruptive game changer on the battlefield (. . .) facilitating extremely flexible 'energy on demand' for combat units and installations (. . .) and providing the ability to deliver rapid and sustainable humanitarian energy to a disaster area".[157] The study concluded that "significant technical challenges remain", that the "best chances for development are likely to occur with U.S. Government support" and that several "key questions/ tasks about SBSP were not answerable within the time of this interim study: identification of clear targets for economic viability in markets of interest (. . .) and full design and deployment of a meaningful demonstrator". Major challenges to be overcome were identified, "including the creation of low-cost space access and a supporting infrastructure system on Earth and in space".

It is obvious that – in addition to the existing substantial technical problems – the major question concerning the economic viability of SPS centres on the mass to be transported into orbit and the corresponding transportation costs. In a DLR-study carried out partly in cooperation with ESA from 1988 to 2001, a new advanced SPS system called 'European Sail Tower SPS' was designed. It was inspired by the 'Sun Tower' concept of NASA but derived and extrapolated from innovative 'Solar Sail' technologies developed in parallel at DLR/ESA.[133,158–161] The structure consists of a 15-km long 'central tether' with 120 power-generating, sun-tracking 'sail modules' attached in pairs. Deployable carbon fibre booms (CFRP) spread quadratic film segments, each 150 m×150 m in size and coated with TFSC. The total mass for the solar array would be 450 tonnes, with a generated power in orbit of 450 MW_e (specific power 830 W_e/kg!). Additionally, the mass for the 1 km microwave antenna was estimated at 1,600 tonnes so that – with very advanced next generation technology – a total SPS mass of ~2,200 tonnes could perhaps be achieved.[131,148] For details, see Fig. 14. The individual 'sail modules' would be brought to LEO for build-up, from where they would travel automatically to GEO with on-board electric thrusters to be

Fig. 14. *Concept of an advanced 'European Sail Tower SPS' in GEO for energy supply on Earth (source: DLR/ESA).*[131,133,148,158,159,161]

integrated into the 'Sail Tower' robotically. After diverse conversion losses, the output on ground would be 275 MW_e, and a very optimistic total specific power on ground of ~125 W_e per kg SPS mass installed in space would result. This is comparable to estimates in other international studies.[146] For a significant contribution to the European or even global electricity supply, thousands of such systems would have to be built and brought into orbit with HLLV and ten thousands of launches. The consequences for the Earth's environment are unknown and might be unacceptable.

A detailed economic assessment of the 'European Sail Tower SPS' was made by Klimke,[131] and the following estimate illustrates the outcome. The electric energy supplied to the grid on ground from such SPS systems during an assumed lifetime of 20 years would be $\sim 2 \times 10^4$ kW_eh per kg installed SPS mass, corresponding to an economic value of about 2,000 euros at energy prices of 0.1 €/kW_eh (roughly the present energy prices, which are also expected for future terrestrial regenerative systems). This energy would only be competitive on the terrestrial energy market including the upcoming regenerative options if development, fabrication, transport into space and operations cost less than about 1,000 €/kg

SPS (necessary ground infrastructure not yet included!). This is not achievable in the foreseeable future, because already the costs for launching (presently 10,000–20,000 €/kg) would have to decrease by more than one order of magnitude.

For long distance power transmission on Earth and as an alternative to terrestrial cables, K. A. Ehricke proposed the following concept in 1972: on the power-generating side on Earth, electricity is fed to a large phased-array antenna on ground using microwave generators to transmit an electronically controlled beam towards an orbital reflector satellite (PRS, preferably in GEO). The PRS intercepts and redirects the power beam to a rectenna at the desired power-consuming location on Earth.[162,163] The idea behind this is to make remote regenerative energy resources accessible (e.g. hydropower in Brazil and Canada, terrestrial solar power in the Earth's deserts) by supplying them to industrialised or developing countries. Long intercontinental transmission distances can justify significant efforts and investment in energy conversion from AC/DC electricity to microwave energy and back since terrestrial competitors like HVDC electricity transmission as well as pipeline and tanker transportation for liquid and gaseous energy carriers also require additional infrastructure (some yet to be developed). For developing countries without an existing extensive electricity infrastructure, PRS may be an alternative for electrification and industrialisation.

PRS are considered to be easier to realise than SPS because of the inherently simpler and smaller design (dimensions $\sim$1 km, mass <1,000 tonnes for $\sim$5 GW_e output) and reduced operational requirements. Constellations with several of these would be required for larger, global systems. Glaser expects the specific costs of PRS to be in the order of 1,000–2,700 \$/kW_e.[162] But this also implies significant decreases in launch costs. Assuming a 20-year operating time (and including costs for maintenance, operation and capital) a few €cts/kW_eh for power transmission might result.

It was also suggested to build large SSP systems not on the Earth's surface but on the Moon and to launch them from there to GEO. Theoretically, this has the benefit of having a much smaller lunar gravitational well, while at the same time taking advantage of lunar resources available for fabrication and launch. Alternatively, one could even leave the solar power plants on the lunar surface and beam the energy down to Earth. David Criswell from the University of Houston promoted this concept of LPS.[164] However, the technological requirements, e.g. the (human) infrastructure on the Moon for manufacture with in situ resources, multiple reflectors (PRS) in GEO to redirect the beams from the Moon to target sites on Earth, etc., as well as the corresponding costs appear exorbitant, with the major necessary technological and operational break-through not within view. Therefore, LPS concepts remain futuristic and are not discussed in more detail here.

5.4.3.2 Terrestrial fusion with ^{3}He from the Moon

Scientists working on the idea of controlled thermonuclear fusion worldwide hope to understand and control the complex plasma-physical processes at temperatures of $\sim 10^8$ °C and more in about 50 years, so that the construction of first power plants can be initiated afterwards. Possible fusion processes with isotopes of hydrogen (deuterium D, tritium T) or helium (^{3}He) are:

$$\begin{aligned} D + D(50\%) &\rightarrow {}^3He + n + 3.3\,MeV \\ (50\%) &\rightarrow T + p + 4.0\,MeV \\ D + T &\rightarrow {}^4He + n + 17.6\,MeV \\ D + {}^3He &\rightarrow {}^4He + p + 18.3\,MeV \end{aligned}$$

Mainly for two reasons, one concentrates presently on the fusion of D + D and D + T respectively:

(1) Due to the single positive nuclear charge of D and T, the repulsive forces between the nuclei are easier to overcome than with higher charge. With ^{3}He, one would have to turn to a more complicated technology with higher plasma-parameter values for the product 'confinement-time × temperature × density' in the "Lawson-criterion" ($\approx 10^{131}$ [s keV/m^3] for D + T and 'break-even' between inserted and gained energy).

(2) D and T are relatively easy to obtain on Earth, while ^{3}He is practically not available in quantities beyond a few kilogrammes. For a significant contribution to the world energy supply, 10–100 tonnes of ^{3}He would be needed per year. The fraction of ^{3}He in the total helium content of the Earth's atmosphere is about 10^{-6} and adds up to a few thousand tonnes, but it is not economically separable. Volcanic vents release a few kilogrammes per year additionally, which are lost again to space in equilibration processes. ^{3}He is also obtained from the radioactive decay of tritium in nuclear reactors or thermonuclear weapons and can be produced artificially in the order of a few 10 kg/year. The world reserves today amount to a few 10 kg.

The disadvantage of D + D and D + T fusion respectively is that major parts of the gained energy emerge as motion of high energy neutrons which must be transformed into usable heat via collisions with the container walls. Thereby, unpleasant radioactive secondary products are formed. If the energy, on the contrary, were transported mainly by charged particles (protons and other nuclei), it could be converted to electricity much easier via interaction with magnetic fields. In this respect, D + ^{3}He fusion would be preferable in principle, although D + D side reactions within the hot plasma also produce secondary neutrons – but much less. Because of the lack of ^3He on Earth, the option of a ^3He supply from the Moon has been under discussion for a fairly long time.[165,166]

^3He together with H, C, N, ^4He and some other volatile elements on the Moon is implanted from the solar wind in extremely small quantities in the lunar regolith (the upper layer on the lunar surface, a few metres thick and broken down to small rocks and fine-grained dust by meteoritic impacts over billions of years). The ^3He content amounts to a mass fraction of typically 10^{-3} to 10^{-2} ppm (parts per million).[167] To extract 1 kg, a regolith volume of 10^5 to 10^6 m^3 would have to be collected and heated to $\sim$1,000°C, and the escaping gases would have to be separated via complex distillation/liquefaction processes (extreme cooling, etc.) with tonnes of hydrogen and other gases accumulated, too. A significant contribution from ^3He fusion to the world energy supply would require an area of about 100 km $\times$ 100 km mined per year with a depth of 1 m! This illustrates the gigantic operating expenses to be carried out on the Moon.

And what would be the economic benefit from lunar ^3He? Via fusion, a thermal energy of 10^8 kWh could be gained from 1 Kg and after conversion to electricity, about 10^7 kW_eh.[168] This corresponds to an energy value of about 1 million euros per kg ^3He (at assumed energy prices of 0.1 €/kW_eh). Although this appears to be a relatively high value compared to most other known products, it remains unclear and doubtful if the costs for the production of 1 Kg on the Moon and transportion to Earth will be so much less as to justify economic mining, keeping in mind also that still unknown costs for the terrestrial fusion power plants must be added. In any case, because of the multitude of unsolved problems – including even fusion technology – a potential utilisation in this century is not expected.

5.4.4. Summary and conclusions

Regarding future sustainable options for the global terrestrial energy supply, Space Solar Power and ^3He from the Moon as fuel for advanced fusion have to be put in contrast to terrestrial alternatives like solar/wind/hydro power, etc. It can be argued that an appropriate terrestrial regenerative energy mix has the potential to satisfy global energy needs. Obviously, these solutions must be taken into account and installed with higher priority within the next decades to reduce the deposition of CO_2 into the atmosphere. Concerning solar power from space, the recently developed concepts for light-weight inflatable and deployable solar arrays/concentrators – like the NASA 'Sun Tower' and the 'European Sail Tower SPS' – have been reviewed and major problems with wireless power transmission have been discussed in this contribution. Compared to earlier concepts, the designs have the potential to significantly reduce the masses and, thus, the costs. But the technological demands and operational uncertainties are still immense. This holds true even more for the lunar ^{3}He concepts. Major progress with cost reduction of one to two orders of magnitude is required for the space options to become competitive with terrestrial regenerative ones. Even though the realisation of the space visions is extremely difficult and not expected in the foreseeable future – both from a technical and an economic point of view – technology development into that direction can be promising, using the visions as a challenge and technology driver (also for other applications in space and on Earth) and to keep options open for the unknown.[150] Anyhow, since the space options are expected to be realisable only in the distant future – if at all – the energy problems concerning supply as well as pollution must be tackled and solved on Earth. To this end, space can clearly contribute with means for observing the Earth, enabling better geophysical prospection, surveillance and the planning of infrastructures.

---

[130] International Energy Agency. World Energy Outlook 2006. Paris: OECD/IEA Publications, 2006. 1 Sept. 2008. http://www.worldenergyoutlook.com/2006.asp

[131] Klimke, Michael. Systemanalytischer Vergleich von erd- und weltraum-gestützten Solarkraftwerken zur Deckung des globalen Energiebedarfs. Diss. University of Stuttgart, Germany, 2000. DLR FB 2001-12.

[132] Hoffert, Martin I., et al. "Advanced Technology Paths to Global Climate Stability: Energy for a Greenhouse Planet." Science 298.5595 (2002): 981–987.

[133] Seboldt, Wolfgang. "Space- and Earth-Based Solar Power for the Growing Energy Needs of Future Generations." Acta Astronautica 55.3–9 (2004) 389–399.

[134] Klimke, Martin and Seboldt, Wolfgang. "Solar Power Plants: Comparison of the Space and Ground Option." Proceedings of the 49th International Astronautical Congress, Melbourne, Australia, 1998. IAF-98-R.2.03.

[135] Spiegel Online. "Strom aus der Sonne: Kosten und Nutzen." 4 Apr. 2007. http://www.spiegel.de/fotostrecke/fotostrecke-20593.html

[136] Melíß, Michael, and Manfred Kleemann. Regenerative Energiequellen. Berlin: Springer, 1993.

[137] Heinloth, Klaus. Energie. Stuttgart: Teubner, 1983.

[138] Bundesministerium für Umwelt, Naturschutz und Reaktorsicherheit. "Entwicklung der erneuerbaren Energien in Deutschland im Jahr 2007: Grafiken und Tabellen." 12 Mar. 2008. http://www.erneuerbare-energien.de/inhalt/39830/20010/

[139] Nitsch, Joachim, et al. Ökologisch optimierter Ausbau der Nutzung erneuerbarer Energien in Deutschland. Forschungsvorhaben im Auftrag des Bundesministeriums für Umwelt, Naturschutz und Reaktorsicherheit, FKZ 90141803. Berlin: BMU, 2004.

[140] Güth, G., Häusler, M., and Schlayer, G. "Übertragung elektrischer Energie über große Entfernungen." Global Link: Interkontinentaler Energieverbund. ed. Verein Deutscher Ingenieure e.V./Gesellschaft Technische Gebäudeausrüstung. VDI Berichte 1129. Düsseldorf: VDI, 1994. 233–249.

[141] Wurster, Reinhold. "Randbedingungen und Systemaspekte für Transport und Speicherung von Wasserstoff." Global Link: Interkontinentaler Energieverbund. ed. Verein Deutscher Ingenieure e.V./Gesellschaft Technische Gebäudeausrüstung. VDI Berichte 1129. Düsseldorf: VDI, 1994. pp. 141–158.

[142] Klimke, Michael. "New Concepts of Terrestrial and Orbital Solar Power Plants for Future European Power Supply." Proceedings of the SPS '97 Conference: Space and Electric Power for Humanity, Montréal, Canada, 24–28 Aug. 1997. Kanata, Ontario: Canadian Aeronautics and Space Institute; Paris: Société des Électriciens et Électroniciens, 1997. pp. 67–72.
[143] Hendriks, C., Geuder, N., Viebahn, P., Steinsiek, F., and Spies, J. Solar Power from Space: European Strategy in the Light of Sustainable Development. Phase 1: Earth and Space Based Power Generation Systems. Collaboration of Ecofys Netherlands, DLR Germany, and EADS Germany by order of the European Space Agency. EEP 03020. Nordwijk: ESA, Nov. 2004.
[144] Summerer, Leopold and Franco Ongaro. "Solar Power from Space: Validation of Options for Europe." Proceedings of the 4th International Conference on Solar Power from Space – SPS '04, together with the 5th International Conference on Wireless Power Transmission – WPT 5, Granada, Spain, 30 June–2 July 2004. ESA SP-567. Nordwijk: ESA, 2004. pp. 17–26.
[145] Glaser, Peter E. "Power from the Sun: Its Future." Science 162.3856 (1968): 856–861.
[146] Mankins, John C. "A Fresh Look at Space Solar Power: New Architectures, Concepts and Technologies." 38th International Astronautical Congress, Turin, Italy, 6–10 Oct. 1997. IAF-97-R.2.03.
[147] Nagatomo, Makoto, Susumu Sasaki, and Yoshihiro Naruo. "Conceptual Study of a Solar Power Satellite, SPS 2000." Proceedings of the 19th International Symposium on Space Technology and Science, Yokohama, Japan, May 1994. pp. 469–476.
[148] Seboldt, Wolfgang, Klimke, Michael, Leipold, M., and Hanowski, M. "European Sail Tower SPS Concept." Acta Astronautica 48.5–12 (2001): 785–792.
[149] Gökalp, Iskender, Calabro, Max, Hollanders Hervé, and Deschamps, Lucien. "Space Solar Energy: A Challenge for the European (and International) Community." 53rd International Astronautical Congress, Houston, TX, 10–19 Oct. 2002. IAC-02-R.1.03.
[150] Mankins, John C. "The Promise and Challenge of Space Solar Power in the 21st Century: Picking Up the Gauntlet." 53rd International Astronautical Congress, Houston, TX, 10–19 Oct. 2002. IAC-02-R.1.02.
[151] Ruth, J., and W. Westphal. Study on European Aspects of Solar Power Satellites. 2 vols. ESA-CR 3705/78/F/DK(SC), 1979.
[152] National Aeronautics and Space Administration. The Final Proceedings of the Solar Power Satellite Program Review, Lincoln, Nebraska, 22–25 Apr. 1980. NASA-TM-84183.
[153] Glaser, Peter E., Davidson, Frank P. and Csigi, Katinka I. Solar Power Satellites: The Emerging Energy Option. New York: Ellis Horwood, 1993.
[154] Canadian Aeronautics and Space Institute; Société des Électriciens et Électroniciens. Space Power Systems: Energy and Space for Humanity. Proceedings of the SPS '97 Conference: Space and Electric Power for Humanity, Montréal, Canada, 24–28 Aug. 1997. Kanata, Ontario: CASI; Paris: SEE, 1997.
[155] Stancati, Michael L., et al. Space Solar Power: A Fresh Look Feasibility Study. Phase 1 Report, SAIC-96/1 038 (Space Applications International Corporation for NASA LeRC Contract NAS3-26565, Schaumburg, Illinois), 1995.
[156] United States House of Representatives, Committee on Appropriations. Bill/Report Making Appropriations for the Departments of Veteran Affairs and Housing and Urban Development. Report 105-00, 25 June 1998.
[157] National Security Space Office. Space-Based Solar Power as an Opportunity for Strategic Security: Phase 0 Architecture Feasibility Study. Report to the Director, National Security Space Office, Interim Assessment. Release 0.1. Washington, D.C.: NSSO, 10 Oct. 2007.
[158] Seboldt, Wolfgang, et al. System Concepts, Architectures and Technologies for Space Exploration and Utilization. SE&U Study, ESA Contract Report 12756/98/NL/JG(SC), 1999.
[159] Seboldt, Wolfgang, Reichert, M., Hanowski, N. and Novara, M. "A Review of the Long-Term Options for Space Exploration and Utilization." ESA Bulletin 101 (2000): 31–39.
[160] Leipold, M., et al. "Solar Sail Technology Development and Demonstration." Acta Astronautica 52.2–6 (2003): 317–326.

[161] Seboldt, Wolfgang, et al. "Ground-Based Demonstration of Solar Sail Technology." Proceedings of the 51st International Astronautical Congress, Rio de Janeiro, Brazil, 2000. IAF-00-S.6.11.
[162] Glaser, Peter. "The Power Relay Satellite." Space Power 13.1–2 (1994): 1–23.
[163] Ehricke, Krafft A. Space and Energy Sources. Seal Beach, CA: Rockwell International Corporation, Space Systems Division, 1992.
[164] Criswell, David R. "Energy Prosperity within the 21st Century and Beyond: Options and the Unique Roles of the Sun and the Moon." Innovative Energy Strategies for CO_2 Stabilization. ed. Watts, Robert G. Cambridge, UK: Cambridge University Press, 2002. 345–410.
[165] Wittenberg, L.J., Santarius, J.F. and Kulcinski, G.L. "Lunar Source of ^3He for Commercial Fusion Power." Fusion Technology 10.2 (1986): 167–178.
[166] Schmitt, H.H. "My Vision of Our Space Future." Space Times 29.3 (1990): 10.
[167] Gibson, E.K. Jr. and Johnson, F.S. "Gas Release Pattern for the Apollo-11 Sample 10086." Proceedings of the 2nd Lunar Science Conference, Houston, TX, 2 Mar. 1971. 1351–1359.
[168] Ruppe, Harry O. "Lunar ^3He for Fusion Energy on Earth?" Acta Astronautica 25.10 (1991) 665.

CHAPTER 6

MOBILITY

6.1 Overview

The context: people are travelling more and more and freight transportation is increasing with globalisation. The management of land, air and sea traffic is a crucial issue with critical economic, safety and security consequences. Sustainable mobility is then paramount and requires a constantly more effective and reliable traffic management in and between those various sectors. At the same time, all modes of transportation impact on the environment to different extents. Aiming at sustainability might therefore require limiting our mobility.

The contribution of space: space-based systems contribute in a unique way to an effective management of mobility. Navigation satellites are essential to, and play a pervasive role in, positioning, navigation and timing applications. Satellite-based systems help increasing the land, air and sea traffic fluidity as well as its safety and security. Combined with navigation systems, space-based telecommunications improve mobility by transmitting vehicles' position information, in particular for ships. In addition to the combined use of navigation and telecommunications, a third kind of space-based system, Earth observation satellites data which provide maps as well as climate data also augment navigation capabilities. The European Galileo satellite navigation system will continue to greatly improve all modes of transport. It will offer a vast spectrum of application possibilities, from maritime applications to precision-farming and emergency services. Moreover, Galileo will include security applications (for instance, crisis management support) as well as sovereignty applications.

The papers: the three papers of this section provide a comprehensive overview of the challenges to be addressed in order to develop sustainable mobility, and the contribution of space-based systems to the necessary increase in safety and security in different transport areas.

In "Threats, Risks and Sustainability – The Challenges to Mobility", Michael Meyer zu Hörste, Working Group Member of the European Rail Research Advisory Council (ERRAC), defines the concept of mobility, reflects on the main challenges to sustainable mobility and identifies the need for the three basic elements of localisation, navigation and communication which are common to all means of transportation. He then successively reviews the challenges to sustainable road, rail and air traffic and summarises the data needed for effective traffic management in both regular and catastrophic cases.

In "How Space Systems Help to Address the Challenges Associated with Sustainability in the Field of Mobility", Mario Caporale from the Italian Space

Agency (ASI) starts with emphasising the increasing demand for mobility and info-mobility, i.e. the combination of processes, services, and applications related to the "intelligent management" of mobility. He then reviews how space systems can contribute to sustainable mobility by increasing safety and security in three sectors, namely civil aviation, maritime navigation, and road and rail transport, and introduces examples of programmes developed at the national level to cope with mobility challenges.

In "Telematic Solutions Based on Space Technology", Kashif Din, Martin Linauer, Jürgen Zajicek and Martin Reinthaler from Arsenal Research present two successfully implemented applications in the field of transport telematics based on space technology. The first application allows a near real-time monitoring of the current traffic situation by using vehicles equipped with positioning devices. The second application is based on the intelligent route adaptation of vehicles carrying hazardous materials.

6.2 Threats, risks and sustainability – the challenges to mobility

Michael Meyer zu Hörste

6.2.1. Introduction: what is mobility?

Mobility is one of the major attributes of the people of the 21st century. Mobility is generally understood to be the movement of people from one location to another. A more comprehensive definition will be used here: mobility means any type of transportation for people or goods via different means of transport. The transportation of goods influences the mobility of people by making use of the same means of transportation and above all, the same infrastructure. Thus, any approach involving issues of mobility, e.g. questions of safety or the quality of service and performance, must take into account both the transportation of goods and that of people. The number of cars as well as the average number of kilometres travelled by people is increasing. Also increasing are the volume of goods transported, as well as the tonnes of goods transported per kilometre (cf. Fig. 1). However, the traffic infrastructure built by man – railways, roads, motorways and inland waterways – is expanding at a slower pace (cf. Fig. 2).

The other "infrastructures" that are not man-made – the airways and the maritime waterways – are not infinite, but they are reaching their point of saturation at the nodes, which are the ports and airports. Their limited capacity is also limiting transport capacity. Modern communications technologies enable the global exchange of information and goods which, contrary to former assumptions, is not declining but rather increasing the demand for the transportation of passengers and goods. So, it can be observed that mobility increases when the economy grows, and an assumption which is made today is that the economy grows as long as mobility increases. Therefore, mobility is a great challenge to sustainability.

The fluidity of traffic must be kept at today's level. In combination with the increasing number of vehicles of all kinds and the slower expansion of infrastructure, this means that there is a need to increase the number of vehicles using one infrastructure unit or, in other words, to reduce the infrastructure required for one vehicle.

The increase of the vehicle density per infrastructure unit and speed on the one hand, and the need for fluid traffic on the other, calls for a higher level of technical

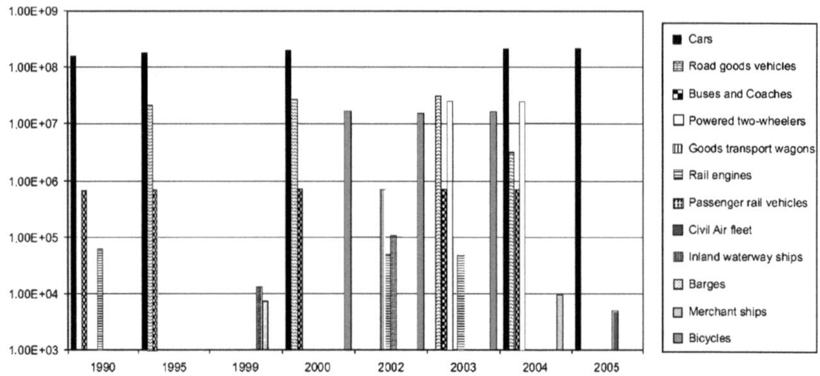

Fig. 1. *Development of number of vehicles from 1990 to 2005 on a logarithmic scale (source: Campbell: European Panorama of Transport).*

reliability and operational safety. The greater volumes of traffic and the on-time requirements of the transportation of people and goods as well as the availability of reserve infrastructures demand high performance, operational reliability, and a good quality of transport services.

Finally, traffic and consequently mobility are responsible for a major part of the impact on the environment. In the year 2004, for instance, 19% of greenhouse gases (GHG) were emitted by means of transport. The transport domain is the only one in the European Union that reported higher emissions (+26%) from 1990 to 2004 (cf. Fig. 3). More than 90% of these emissions originated from road transport.[169]

There is an unquestioned need for research in order to find a response to the challenges mentioned above. The European Commission has therefore launched

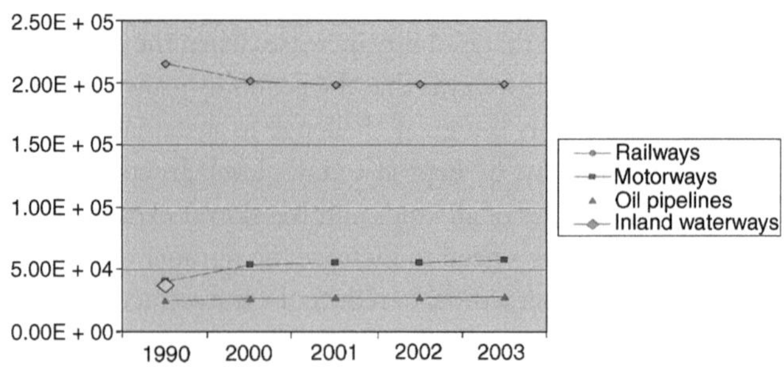

Fig. 2. *Development of transport infrastructures from 1990 to 2003 on a logarithmic scale (source: Campbell: European Panorama of Transport).*

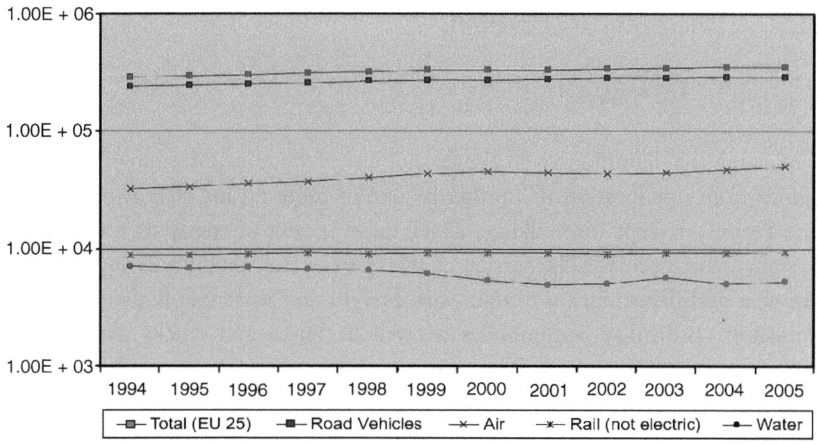

Fig. 3. *Energy consumption of transport in 1000 tonne oil equivalent on a logarithmic scale (source: Campbell: European Panorama of Transport).*

so-called European Technology Platforms (ETP). These ETPs are groups of stakeholders for a specific domain. All stakeholders can join and the lead is normally taken by an industry representative. They define research and development priorities as well as timeframes and action plans on strategically important issues for the domain. The results are compiled in a document called "Strategic Research Agenda" (SRA). This document is used by the European Commission to focalise the content of the research framework programme – mainly the Seventh Research Framework Programme – so as to meet the needs of the industry and operators in Europe. Currently, more than 30 ETPs are active.[170] For the domain of transport, four ETPs are active for the aeronautics, road, rail, and waterborne sectors. There is another ETP for space technology that influences the domain of transport as well as many other technology-oriented ETPs that will not be discussed here. The main research topics for the domains of aeronautics, road and rail are presented in the following sections and refer to the relevant SRAs.

The next section contains a summary on the topic of mobility in general. The following three sections focus on the domains of road, rail and aeronautics, respectively, these being the most important means of transportation. Section 2.6 focuses on traffic management as a generic function of all modes of transport. A short, last section on space transport discusses potential means of transport in the future. Each of these sections gives an overview of the specific domain and discusses open aspects as well as the relevant space technology already in use.

6.2.2. Mobility in general

6.2.2.1. Definition

Following the definition given above, mobility is the transportation of persons or goods from one location to another by one or more means of transportation in most cases. Except for walking, all of these means of transport are technical systems which provide the service of transport. Thus, mobility is supplied by a number of different means of transport. Freight and passenger flights provide air transport. Individual automobiles as well as buses and trucks use the road network and the highways. Tramways, metros, mass transit as well as regional and long distance railroads provide railway transport. Ferries and freight ships use inland waterways. Goods and passengers are transported by maritime ships. Even walking or the use of bikes, escalators, or funiculars are types of mobility. The most futuristic types are space transport for commercial satellites and space tourists.

An important aspect of mobility is the differentiation between individual and public transport. Individual transport uses primarily automobiles and motorbikes for motorised transport. Private airplanes, ships and others means of transport are not so significant for the volumes of traffic. Public transport is provided by cars, buses, tramways, railways, airplanes and ships of different kinds.

6.2.2.2. Generic challenges of mobility

Some aspects are invariant for any mode of transport: the persons in charge of control of the vehicle, e.g. the drivers, pilots and captains need to know where the vehicle is, what the current movement status is, and in which direction it has to move in the immediate future as well as how to reach the point of destination. The first two aspects are subsumed in localisation, and the latter two constitute navigation. The person in charge has to know the status of the traffic and infrastructure in front of the vehicle. The driver either sees the situation or has support from sensors in the vehicle or is informed by the control centre. The safety as well as the security of the transport must be guaranteed. Transports of goods – especially hazardous goods – have to be tracked and traced carefully. The impact on the environment should be minimised. Some of the technological aspects are similar for many modes of transport:

(1) Locating: at least for autonomous control, the vehicle needs to know where it is. As soon as traffic management is involved, this information needs to be transmitted to the infrastructure or to other vehicles. Generally, there are

different possibilities for locating a vehicle: the vehicle can identify its position itself or a control centre reports the location of the vehicle.

(2) Navigation: navigation is a process of planning and guiding the movement of the vehicle from one place to another.

(3) Communication: as soon as more than one vehicle uses the area covered by the same control centre, there is a need to exchange information. As a minimum requirement, this information will comprise the position and current movement of the vehicles.

There are different ways for implementing these technological aspects.

6.2.2.3. Traffic management

Practically the entire system of public transport and an increasing share of individual transport are coordinated by traffic management centres. These centres operate very differently for the various means of transport. For railways and airports, these centres carry most responsibility for operational safety. Traffic management centres for road traffic just distribute status information and give redirection advice to drivers.

Normally, traffic management operates smoothly and more or less undisturbed. For optimisation purposes, standard data are collected such as the number of vehicles operated on a certain piece of infrastructure, punctuality, average speed, etc. The traffic situation as well as the data change fundamentally if there is a massive disruption of the traffic situation. A major accident or a disaster can yield such a result, but traffic in the case of a public mass event is also very similar in this regard. Section 2.6 of this article focuses on the aspects and challenges of traffic management.

6.2.3. Road traffic

Road transport is the most important mode of transport and has been increasing steadily over the past years. Freight transport on roads in Europe had a share of 44% in 2005, this being an increase of almost 38% since 1995. Passenger transport accounted for a share of more than 75% in 2004, with an increase of around 18% having occurred since 1995. In 2005, there was nearly one passenger car per two inhabitants in Europe. From 1990 to 2003, the network of motorways grew by 41%.[171] Consequently, research on road transport as well as mobility using roadways is one area of focus of European research politics. The European Technology Platform for road transport is the European Road Transport Research Advisory Council (ERTRAC). The need for research and development is described in the strategic research agenda (SRA)[172] that was developed by ERTRAC. The following goals are stated in the SRA:

- Mobility, transport and infrastructure
  - Mobility of people
  - Transport of goods
- Safety and security
  - Accident prevention
  - Accident impact mitigation
  - Road transport system security
- Environment, energy, resources
  - Reduced GHG emissions and more efficient energy use
  - Environment, including impact on communities and natural habitats
- Design and production
  - Time to market and implement
  - Flexible production systems
  - Lifetime resource use.

Some major technological challenges to road transport are:

Today's cars support drivers in their driving tasks to improve traffic safety and fluidity. A driver who is aware of the situation and well informed about the traffic status knows what he or she has to do to drive safest and take the best route. Advanced Driver Assistance Systems (ADAS) are implemented in cars to improve situational awareness and to supply information. The first generation of driver assistance systems had a defined, fixed functionality. Once they were activated, they always worked the way the driver had set them. The newer and much more interactive generation of ADAS act commensurate with the driver, the traffic situation and the destination of the driver which is normally a specific location. The typical attributes of a driver – which have to be taken into account by the ADAS – are, for example, gender, age, and driving experience.

Nearly all ADAS need to know where the car is. This is the classical aspect of localisation: where is the car and how fast is it moving? Today, the localisation platforms contain an algorithm for tracking the vehicle using a digital map. Current systems are intelligent enough to track a car in a city without using GPS. Nonetheless, localisation is a field for further improvement.

Another requirement of ADAS is information: static information about road networks, vehicle characteristics, speed and weight limits, traffic rules, as well as dynamic information about the traffic situation, disturbances and accidents. This information is delivered by wireless telecommunications equipment. In the past, this type of communication was between the vehicle and the control centre infrastructure. Currently, many concepts are based on vehicle-to-vehicle communication, so-called Car2Car communication.[173]

Another important aspect is the safety and reliability of vehicle control. The ADAS are developing from providers of purely supportive and advisory functions to intervening and/or controlling functions. Thus, the technological safety of the vehicle as well as its operational safety depends on the correct operation of the systems.

6.2.4. Railways

Railways are a major means of public transport. The mobility aspects related to railways can be separated into two main fields: first, the provision of information to customers, i.e. travelling persons or freight operators. Second, railway control and operation management. Research concerning railways is another important field of European research politics. The European Technology Platform for railways is the European Rail Research Advisory Council (ERRAC). The need for the development of the railway system is described in the strategic rail research agenda (SRRA)[174] which was developed by ERRAC. The following aims are stated in the SRRA:

- Interoperability
  - Future traffic and transport management systems for mainline and urban transit – research, migration and implementation strategies
  - Technologies to facilitate dedicated freight lines
  - Toolkit to enhance interoperability and modularity
  - Common operational rules
- Intelligent mobility
  - Seamless transport (passenger and freight)
  - Harmonise existing and/or new services for operators (passenger and freight)
  - Harmonise existing and/or new services for freight customers relations
- Safety and security
  - Safety and security philosophy and culture
  - Human safety and security factors
  - Assuring customer safety and security
  - Safety and security relevant structures of information technologies
  - Modelling and simulation
- Environment
  - Improving noise performance of the railway mode
  - Reduce emissions to the environment

  - Reduce operational energy consumption
  - Design for the environment
  - Development and deployment of emerging technologies
- Innovative materials and production methods.

In both of the fields mentioned above, one of the most important questions is: where is the train? This includes information about the driving status as well as the identity of the train. The major difference is that the customer typically requires information with a lower degree of precision and without safety relevance. The train control centre, in contrast, requires precise information with a high degree of precision of about 1 m with absolute safety relevance. The major technological challenges for rail transport are discussed in the following section.

Railways are one of the safest modes of transport. But there are different approaches to ensuring safety in the different countries in Europe. The establishment of a harmonised European concept for ensuring safety is one of the major challenges for railways. Besides, railways are also one of the most environmental-friendly modes of transport. Less than 1% of the transport-related emissions of GHG come from railway engines and traction units.[175] Yet there is a need to reduce the emissions of non-electrical traction units.

6.2.4.1. Customer information

The mobility of passengers in public transport is based on a specific kind of navigation task. The planning of a journey is carried out on the basis of many different databases containing routes, schedules, timetables, and connections. During the journey, this planning must be supervised and – if necessary – modified. Typically, this type of supervision is based on a kind of time-position relation. Thus, the question that arises is whether there is a difference between the position where the vehicle has to be at a given time, and the real position of the vehicle at that time?

The technological challenge concerning mobility is to be able to support passengers by supplying the right information at the right moment. A passenger needs to know whether the journey planned can be carried out or if re-planning is necessary, and which of the planning and current information is of relevance. The basic inputs for this process are the current time and position of the train including the identification of the train. Therefore, the basic technological needs of localisation, navigation and communication are relevant for railway passengers as well, but with different characteristics.

The issues for freight customers are similar. The main question is whether the transported goods will arrive on time or – if there is no pre-defined arrival time –

the estimated time of arrival. To estimate the time of arrival, the current position of the transport is used. It is already state-of-the-art to supervise the transportation of a good. The customer as well as the freight forwarder can track the position of the freight unit via a specific, secured Internet site and also obtain additional related information.

6.2.4.2. Railway operation management

Railway operation management has the basic task of ensuring the safe operation at defined levels of quality. Therefore, trains, vehicles, transport units and transported items must be tracked and traced. Normally, operation management is carried out on the basis of pre-planned data: timetables, vehicles in circulation and roster, etc. These tasks are already highly automated today. A higher need for the interaction of the operators arises in situations of failures, accidents, or construction work.

What is required in this context is the improvement of the performance and cost-efficiency of railway systems. On lines with high traffic density and/or high speeds, it is necessary to run trains with shorter headways to increase the capacity of the lines. On lines with medium or low traffic density, it is necessary to reduce the cost of the control systems, specifically in the areas of cables, sensors, axle counters, track circuits, etc. Moreover, in situations of failures, construction work or accidents, it is also required to keep the railway system operational as far as possible.

Especially railway networks with very low-density traffic require very cost-efficient train control solutions. Approaches involving a Car2Car environment are being discussed, e.g. the idea of a Railway Collision Avoidance System[176] to reduce the control system at the infrastructure centre as far as possible and to install all of the required technology in the trains.

6.2.4.3. European Train Control System (ETCS)

The generic questions for all kinds of train control systems are: where is the train, including its identification, and which is the guaranteed safe part of the track? Historically, each country had a nationally defined train control system. For nearly 20 years, work has been done to specify and implement a harmonised ETCS as one of the main components of the European Rail Transport Management System (ERTMS) which would ensure the interoperability of trains in Europe. A specific challenge of railway-related mobility is the transfer of the train control system towards the ETCS. In the next few years, the first implementations of the ETCS will begin to operate. For example, in Spain on the line from Madrid to Barcelona, there will be on-board units from different suppliers that must work interoperably with trackside sub-systems from other suppliers. To avoid additional investments of time and money, the interoperability of these components will be tested in a lab before testing and operating them on the track. Different tests must be performed: technical safety tests and reliability and availability tests on the one hand, and operational safety tests and conformity with specifications tests on the other hand. One very challenging issue at the technological level of the introduction of the ETCS is to find a good solution for the precise, reliable and safe localisation of the trains. The application of the European navigation satellite system Galileo has been studied in detail as one possible solution for this issue. The first results were published on the Internet by the European research project "GNSS introduction in the Rail Sector" (GRAIL)[177] funded by the GNSS supervisory authority (GSA).

6.2.5. Air transport

In the past few years, the sector of air transport has grown very fast. The ratio of passengers per kilometre has grown by nearly 50% between 1995 and 2004. The tonne-to-kilometre ratio went up by more than 30% during the same period.[178] As a result, aeronautics research has become a well-established field of European research politics. Similar to the fields of road and rail transport, the Advisory Council for Aeronautics Research in Europe (ACARE) has developed a Strategic Research Agenda 2.[179] The following goals are identified in the SRA2:

- Quality and affordability
  - Reducing travel charges
  - Increasing passenger choice
  - Transforming air freight services
  - Creating a competitive supply chain able to halve time-to-market
- Environment
  - To reduce fuel consumption and CO_2 emissions by 50%
  - To reduce perceived external noise by 50%
  - To reduce NO_x by 80%
  - To make substantial progress in reducing the environmental impact of the manufacture, maintenance and disposal of aircraft and related products
- Safety
  - Reduction of the accident rate by 80%
  - Reduction in human error and its consequences

- Air Transport System efficiency
  - To enable the Air Transport System to accommodate three times more aircraft movements by 2020 versus 2000
  - To reduce the time spent by passengers in airports to under 15 min for short-haul flights and to under 30 min for long-haul flights
  - To enable 99% of flights to arrive and depart within 15 min of their advertised scheduled departure time, in all weather conditions
- Security
  - Zero successful hijacks.

The major challenge for air transport is the massive increase in transport demand. This requires the following actions:

(a) Development of the infrastructure at airports and the further development of air traffic management to cope with the dramatic increase in flights.
(b) Expansion of airplane-holding capacities as well as procedures for checking safety and security.
(c) Improvement of passenger-handling capacities as well as freight-handling capacities.

A resulting important issue is the reduction of emissions and noise to reduce the impact of flights on the environment.

6.2.6. Fields of research: traffic management

As already discussed in the introduction, most of the means of transport interact with traffic management. Normally, these traffic management centres serve only one mode of transport. Sometimes, integrated traffic management is used for all ground-based public means of transport, e.g. railways, tramways, buses and metros. In normal situations, traffic management works on the basis of pre-planned data to optimise the fluidity of traffic. In the case of a major accident, a catastrophe or public mass event, traffic management works differently. In this case, enabling traffic and the prioritisation of qualified traffic participants is the main objective. Both cases are discussed in the following section.

6.2.6.1. Under normal circumstances

In a normal situation, the task of traffic management is to ensure the performance as well as the safety of traffic. Within the domain of road transport, traffic management is based on statistical variables such as traffic density, average speed, etc. For maritime, air and rail transport, traffic management is based on individual vehicles. The operation of traffic management is based on the following data:

- Topology of the infrastructure: road maps, track databases, etc.
- Schedules or timetables
- Identity tables, vehicles in circulation and roster plans.

These data are typically available in advance. Ad hoc scheduling is based on differences to the existing planning. The aim of traffic management is to optimise the performance and fluidity of traffic while keeping it safe. Some major technological challenges for traffic management are discussed here. Traffic and vehicle flows must be measured and the data must be transmitted to the traffic management centre. Measures to influence traffic must be transmitted in the other direction to the traffic infrastructure or vehicles.

6.2.6.2. ... And in the case of a disaster or public mass event

The situation of traffic management changes massively as soon as the situation is heavily disrupted. This can happen momentarily when a disaster or major accident occurs. A very similar traffic situation results from public mass events. In these situations, the availability of data is reduced. The planning data are usually no longer valid, even if there has been a mix of forecasting and planning for a public mass event. The following questions are typical examples: is the traffic infrastructure still or again available, usable, and not saturated? Which traffic flows exist? How fast are they flowing? How can priority transports (ambulances, search-and-rescue teams, police, fire-fighters, etc.) pass by the other traffic?

The detection, identification, and measurement of the strength and direction of traffic flows are major tasks in the event of a disaster. Local traffic sensors and measurement units are typically unusable unlike in the normal situation, because they do not exist, are not operational or no longer trustworthy, or possibly do not work outside the usual information infrastructure. Remote sensing technologies help to obtain the required information. The on-demand distribution of the required information to qualified receivers is the next step in the traffic management process. Robust and flexible communication networks with a minimum of local infrastructure are needed in this case. Finally, localisation technologies are required to locate the different teams and emergency cars.

Table 1 shows the major characteristics of both types of traffic management.

Tab. 1: *Traffic management.*

	Regular case	Catastrophic case
Detection of traffic flows	Strength and direction	Detection, identification, strength and direction
Main characteristic of traffic flows	Fluidity	Existence
Planning and forecast of traffic, based on	Maps, schedules, local sensors	Remote sensing
Distribution of information	Common information technologies	Secured
Information distribution technology	Broadcast	On demand
Information receiver	All	Only qualified
Selection of relevance	On receiver side	On sender side

6.2.7. Other modes of transport

6.2.7.1. Waterborne transport

The Waterborne European Technology Platform is the fourth Technology Platform that has defined a strategic research agenda[180] concerning the demand for research and development in that field. The technological challenges are similar to the challenges related to the other means of transport: vessels need to be located, identified, and information has to be transmitted by means of communication.

6.2.7.2. Space transport

Transports by rockets into space were historically only performed for military, public or scientific purposes. In the past few years, however, there have been a significant number of space transports for commercial purposes (cf. Fig. 4). Normally, such commercial space transports are performed for satellites. Space tourism is the only means of transport for which the names of all passengers are available on the Internet.[181] There are forecasts which state that this kind of transport will strongly increase in the future.[182] The European Technology Platform for space technology is the European Space Technology Platform (ESTP) which has published a SRA addressing the most important topics of European Space Research.[183]

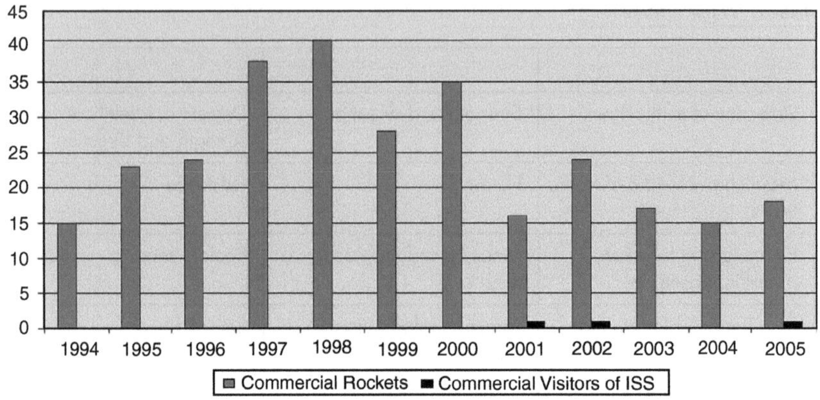

Fig. 4. *Evolution of commercial space transportation from 1994 to 2005 (source: Eurostat).*

6.2.8. Conclusions and outlook

Mobility is a major necessity for people in the 21st century. Mobility is growing fast, massively influences the environment, and also accounts for a major part of all fatalities caused by traffic accidents. As a result, research and development for sustainable mobility constitute one major aspect of research politics in Europe.

Some technological aspects are relevant for almost all modes of transport. Vehicles need to be located to allow for assistance, supervision, and control. As soon as more than one vehicle uses a similar infrastructural area, the need for communicating data arises.

The impact of all modes of transport on the environment as well as the number of fatalities needs to be reduced. A well-designed traffic management system can help to improve the global traffic situation, but what is most needed is the integration of the different modes of transport.

---

[169] European Commission; Eurostat; and Glen Campbell. Panorama of Transport, Edition 2007. Luxembourg: Office for Official Publications of the European Union, 2007.

[170] CORDIS. "European Technology Platforms (ETPs): Individual Platforms." 14 Feb. 2008. http://cordis.europa.eu/technology-platforms/individual_en.html

[171] European Commission; Eurostat; and Glen Campbell, 2007 (see endnote 1).

[172] European Road Transport Research Advisory Council (ERTRAC). Strategic Research Agenda. December 2004. http://www.ertrac.org/pdf/publications/ertrac_agenda_dec2004.pdf

[173] See: Homepage of the Car2Car Communication Consortium. http://www.car-to-car.org/

[174] European Rail Research Advisory Council (ERRAC). Strategic Rail Research Agenda 2020. September 2002. http://www.errac.org/docs/ERRAC_SRRA_Official.pdf

[175] European Commission; Eurostat; and Glen Campbell, 2007 (see endnote 1).

176 See: Homepage of the Railway Collision Avoidance System (RCAS). http://www.collision-avoidance.org/rcas/index.php?lang=de

177 GRAIL Project. "GNSS Introduction in the Rail Sector." 14 Feb. 2008. http://www.grail-project.com/GrailWEB/htm/home.htm

178 European Commission; Eurostat; and Glen Campbell, 2007 (see endnote 1).

179 Advisory Council for Aeronautics Research in Europe (ACARE). Strategic Research Agenda, Vol. 1. October 2004. http://www.acare4europe.org/docs/ASD-volume1-2nd-final-ss%20illus-171104-out-asd.pdf.

180 Waterborne European Technology Platform. Strategic Research Agenda Overview: Waterborne Transport & Operations, Key for Europe's Development and Future. May 2006. http://www.waterborne-tp.org/download_document.phtml?file_id = NzMyY2Ju=

181 Wikipedia. "Space Tourism." 14 Feb. 2008. http://en.wikipedia.org/wiki/Space_tourism#List_of_flown_space_tourists

182 See: Homepage of Virgin Galactic. http://www.virgingalactic.com/htmlsite/introduction.php

183 European Space Technology Platform. Strategic Research Agenda. June 2006. http://estp.esa.int/upload/ESTP_SRA_1.0-July2006.pdf

6.3 How space systems help to address the challenges associated with sustainability in the field of mobility

Mario Caporale

6.3.1. Introduction

Today, demand for mobility is growing. People are travelling more and there is a dramatic increase in freight transport due to the increased circulation of goods at a global level. This increase in mobility implies new safety and security issues. In addition, people demand more and more information in the area of mobility. Technological advances are driving a growing demand. Specifically, space systems can help to satisfy the rising demand for mobility – info-mobility – that is secure and safe.

Info-mobility is the combination of processes, services, and applications related to the "intelligent management" of mobility. It is a key factor for economic and technological growth. Info-mobility includes so-called LBS (Location-Based Services), i.e. the provision of services and information to users based on their mobility and geographical position. Space positioning and location is already pervading human life in many areas, but the most evident is the automotive sector. GPS (Global Positioning System) is increasingly being used in private cars to support personal navigation.

International initiatives have developed special augmentation systems (SBAS: Space Based Augmentation Systems) such as EGNOS (European Geostationary Navigation Overlay Service) in Europe, WAAS (Wide Area Augmentation System) in the U.S., and MSAS (Multifunctional Satellite Augmentation System) in Japan to provide reliable navigation signals for civil aviation, maritime navigation, rail and road transport that ensure the safety of life.

The American GPS system and the Russian GLONASS (Global Navigation Satellite System) were created mainly for military purposes, but their use, especially in the case of GPS, has spread to the civil sector. Due to some limitations in the current GPS signal, there is a lack of integrity regarding the information content of the signals. And because there is no guarantee for the quality of the signal, the accuracy is not always sufficient for many civilian applications. Furthermore, with worldwide coverage being not good especially at high latitudes, Europe has decided to develop a new system called "Galileo", which will mainly be used for civilian purposes and is specifically conceived to support safety-of-life navigation.

Space telecommunications already support deep-sea navigation by enabling the continuous exchange of information and tracking control. The AIS (Automatic Identification System) is supported even in a deep-sea environment via satellite navigation and telecommunications systems.

Earth observation from space provides an ever-increasing supply of detailed information to support navigation with high-resolution mapping on ground, information on waves and winds on seas, weather information, as well as feeding data base systems, which are capable of providing end users with very useful information on mobility.

This paper addresses three sectors in which space systems can be of great help in sustaining mobility in the near future, also with respect to safety and security:

(a) Civil aviation
(b) Maritime navigation
(c) Road and rail transport.

The Italian Space Agency (ASI) promotes the development of application projects in the field of maritime, aeronautical, and terrestrial transport. These projects focus mainly on safety aspects and aim to contribute to increasing the efficiency of traffic management infrastructures by the use of navigation satellite technology, in particular Galileo, EGNOS, and modernised GPS. The scope of work is also paving the way for the advent of Galileo, and applications with this outlook are being developed to prepare the national infrastructures for fully benefiting from Galileo services.

Innovative solutions for applications are defined and developed in these projects; they are technically and operationally validated by means of the involvement of users. The projects motivate companies to invest in the creation of new technologies and products, provide benefits to the overall Italian economy, and open new possibilities to attract private investors to this challenging sector. Research and innovation are promoted while creating public benefits and meeting user needs.

6.3.2. Civil aviation

The European civil aviation sector is currently undergoing a deep transformation in line with the strategy of European integration and the so-called

"Single European Sky". Fundamental to this process is the progressive introduction of satellite navigation into the civil aviation procedures.

6.3.2.1. En-route navigation

Safety and security in air traffic call for instruments and procedures that enable adequate responses to threats to the plane, the crew, or the passengers. Three "main" services, which are briefly described here, are considered essential.

Interference monitoring checks for interferences and reduces their effects; moreover, it reports data on interferences to the ground service centre which are then broadcasted to all aircrafts in the same area. This information can be exchanged via a satellite communications system (i.e. SDLS – Satellite Data Link System).

A radio-assisted function offers periodical reporting that includes the aircraft's position. This service could be provided in areas in which terrestrial infrastructure (i.e. VOR – VHF Omnidirectional Range) cannot assure adequate support.

The consistency of a flight plan allows the continuous checking if the aircraft route is compliant with the route memorised in the Flight Management System (FMS) mirrored in the service centre. If it is not compliant, the ground service centre notifies the aircraft crew and, if necessary, informs – by a dedicated procedure – the air traffic organisations and/or the relevant security institutions.

To carry out the functions described above, the following physical segments of the system are required:

- A mobile segment, consisting mainly of the Aeronautical Earth Station (AES – an SDLS component), the GNSS receiver, the FMS and the on-board computer;
- A space segment (GPS, SDLS, EGNOS, and Galileo);
- A ground segment, mainly composed of the Ground Earth Station (GES – an SDLS component), the WAN (Wide Area Network) and the server.

6.3.2.2. Approaching and landing

Today, within the EU, 15% of all delayed flights are caused by congested traffic to the terminal areas and 9% by adverse weather conditions. The further increasing air traffic projected for the next few years calls for new procedures and advanced instruments for airport approach and landing. In this context, the possibility of approaching the airport using precision procedures (standard P-RNAV – Precision-Radio Navigation) and to drive the plane automatically at the beginning of the airstrip permits landing even under heavily adverse weather conditions. Three services are particularly relevant in this context:

The GNSS Terminal Area Navigation-Precision Approach processes on-board navigation parameters of the aircraft according to the requested performance and with respect to the assigned approach path. In addition, it allows the interfacing of on-board instruments in order to visualise the computed information and receive information from the ground (service centre) to implement the above functions.

An information broadcast capability allows the defining of corrections applicable to the airport area in order to augment the position accuracy and send it to the users. In addition, it allows the monitoring of interferences and the sending of relevant information to the users. It enables an interfacing approach with the control centre in order to receive information linked to the approach path.

A Virtual Reference Station Service enables the supply of correction services based on the information provided by a network of local elements distributed in the area (also to areas that are out of the range of the coverage of the Information Broadcast).

6.3.2.3. At the airport

At the airport, there are a number of additional vehicles which provide services for flights: support vehicles and personnel vehicles, passenger transfers from the terminals to the airplanes, baggage transfers, and safety service personnel transport (firemen, civil protection and health personnel). There is a growing demand for safe movement in the airport area to avoid the runway incursions which can be the cause of disasters. There is a growing need to have a system dedicated to the simplification of traffic management at airports.

Such a system could be conceived to cover specific user needs that are not already implemented in the available systems by applying satellite navigation technologies and, at the same time, ensuring an interface and cooperation with the existing systems and procedures, in particular with the A-SMGCS (Advanced Surface Movement Guidance and Control System). This system could be comprised of three components: a mobile component, a service centre, a local element and a space segment (GPS, Galileo, and EGNOS).

Italy is progressively introducing the new technologies, in particular satellite navigation, into civil aviation and is developing a national initiative to capitalise on the previous investments in programmes such as MTB (Mediterranean Test Bed), MEDUP (Mediterranean Upgrade Programme), MFF (Mediterranean Free Flight) and STENAV (Satellite Test Bed for Navigation and Communications). Italy has invested significant sums in EGNOS and Galileo and is working to achieve the effective exploitation of these two systems. The initiative includes not only the incorporation of EGNOS signals into air traffic control procedures, but also the experimentation with future Galileo signals for take-off and landing, for airport surface vehicle movement management, and for the introduction of satellite navigation to general aviation, helicopter navigation, and so-called UAV (Unmanned Aircraft Vehicle) navigation.

6.3.3. Maritime navigation

There are two areas of maritime transport which deserve attention and in which Italy has concentrated its research activities for satellite navigation applications: maritime highways and personal navigation. Within these areas, two applications have been developed for which innovative solutions based on positioning satellite use have been defined and tested. End users that would benefit from the development of applications in these areas are companies developing industrial applications, port authorities, maritime transportation companies and coast guards, as well as non-profit maritime users.

6.3.3.1. Maritime highways

Trade transport is conducted under conditions of insecurity. Goods travel mainly by road, i.e. by truck and vans that congest motorways and create repercussions of different kinds. Economically, road transport entails delays in delivery and problems with planning and production. Socially, road transport causes anxiety and is harmful to people.

Environmentally, it causes energy waste, toxic emissions and a high impact on common infrastructures.

The way to freeing roads from heavy traffic is to promote inter-modality, i.e. the matching of all possible modes of transport (rail, road, sea, air) with the goal of optimising resources, materials and energy. The key feature of an inter-modal system is the junction-point where it is possible to stop and change the transport modality, and it is important that everything runs efficiently and well-organised to avoid delays which affect the whole transport system.

Maritime transport, if sufficiently supported, is part of an inter-modal system, especially since Italy has been provided by nature with thousands of kilometres of coastline. Transport systems slow down near borders and therefore, the sea – as an infrastructure which does not share this weakness – is not exploited enough.

Guided navigation and berths provide functions and services for managing traffic in the surroundings of harbours and will positively affect seaport systems thanks to greater accuracy, safety and speed in entering and mooring, thereby ensuring the effective planning of seaport activities and the keeping of arrival schedules.

Essential for guided navigation is an integral satellite datum. This is why an augmentation system has been developed which largely uses Galileo services to reinforce the power of a satellite navigation system (accuracy and integrity). This system provides information on the differential correction of satellite signals and measures its integrity in order to estimate positions within a tolerance range of a few centimetres and to comply with the precision criteria of the IMO (International Maritime Organization) standard.

The main services provided by the satellite navigation application developed in Italy are: support for approach operations, additional functionalities to manage maritime traffic near and inside the port, and interferences management to manage the interferences on GNSS SIS and increase the robustness of the system.

6.3.3.2. Maritime personal navigation

Ninety-two percent of the rescue missions of the Coast Guard concern leisure yachts of which there are about 800,000 in Italy, 73,311 of which are registered (UCINA [Unione Nazionale Cantieri e Industrie Nautiche e Affini] 2006). Many of these missions are caused by adverse weather situations and rough seas. The distress calls coming from the yachts in trouble via VHF or mobile phone are often incomplete and the information is not enough to build a proper scenario and organise an effective and fast rescue. The prevention of risks and assistance for rescues is the main contribution of the new technologies. This enhances the popularity of leisure sailing and reduces the work of the Coast Guard.

There are three key moments in leisure sailing: the ordinary situation, the potential warning, and the emergency/rescue situation. During normal navigation, information is supplied to the sailors from Meteo forecasts and ship traffic so that they are able to chart the best navigation course. Potential risk is the second scenario. The service centre can detect, also in the absence of a distress call, if a unit is in trouble because of adverse weather, if it is in the way of fast ships, or if the route changes suddenly.

Finally, in the case of an explicit distress call, any available information (kind of boat, crew, situation, injured aboard, etc.) is provided to the Coast Guard which can then use the right quality and quantity of resources and save resources for other rescues.

The services designed to support personal navigation are based mainly on the new GNSS technologies like Galileo that help to foresee risks and improve rescues, and on VTS (Vessel Traffic System) technology that enables the avoidance of the risk of collision with other ships that are bigger and faster nowadays than in the past (a fast ferry of 11,000 dwt and 145 m can exceed 40 knots).

6.3.4. Road and rail transport

Mobility on the Earth's surface poses different challenges depending on the various modalities of transport. Let us consider two major cases of surface mobility: rail and road. Regarding road transport in particular, let us focus our attention on the challenging transport of hazardous goods.

The main goals of surface info-mobility are:

- To minimise the uncertainty of transfer times, e.g. in unforeseen traffic situations;
- To provide users with suitable emergency assistance;
- Assistance in urban and extra-urban areas;
- To customise end-user services through special-purpose menus and options which define user profiles;
- To allow the possibility to plan personal mobility;
- Mobile computing which allows various services such as electronic toll payment, pay-per-use associated to motorways, parking, controlled access, etc.

6.3.4.1. Road transport: the case of Dangerous Goods Transportation (DGT)

The transportation of dangerous goods is a challenging aspect which can significantly benefit from the use of new technologies, especially satellite navigation and communication. While representing only a fraction of goods transported, dangerous materials contribute disproportionately to transport-related safety issues. Accidents have happened with consequences for human life, the environment and economic infrastructures.

The transport of dangerous goods has a heavy impact on overall freight transport, especially by road: Some European statistics highlight that it accounts for 10–20% of the overall goods transport. New technologies, e.g. satellite navigation, broadband telecommunication, and GIS systems can provide significant support for the management of the various phases of dangerous goods transport: from knowledge and prevention to emergency management and restoration.

Italy is developing an innovative technological platform for monitoring and managing dangerous goods transport and for support in emergency situations after an accident. The system is based on emerging technologies and includes a full set of tools to support the transport of dangerous goods in all transport phases. It is based on GNSS technologies, in particular on the use of Galileo and EGNOS/GPS, and defines not only innovative inter-vehicular communication to support the sustainability of the application, but also a local segment for local augmentation.

The added value of the system lies in a substantial process innovation in the field of emergency and rescue management through the creation of a telematics infrastructure enabling the various emergency management agencies to coordinate their efforts. The aim of the project is in fact to provide these agencies with a communication system and a shared information record, reducing to a minimum the consequences of any risk situation and possibly to even prevent those risk events through a set of risk-reduction control actions. The project has also been developed taking into account the principles provided by ARTIST (Italian Telematic Architecture for Transport Systems) and with a view to creating a modular and open architecture.

The main system capabilities provided are:

(1) *Extended precision localisation:* like other satellite systems such as GPS, the Galileo system ensures a localisation accuracy of up to a few metres. This would be a limitation for functions like monitoring DGT operators' driving behaviour unless augmentation techniques (i.e. methods to increase the precision level of a localisation measurement) were applied. The available augmentation techniques for the GNSS satellite system are the extensions provided by the EGNOS systems and by the Real Time Kinematics (RTK) system, which provides the greatest precision (about one centimetre). However, the use of this technique requires the installation of RTK stations whose corrections provide coverage for a specific limited area. In order to overcome the RTK service coverage problem, the system uses other complementary techniques for increasing precision such as the EGNOS augmentation system that works with GPS, and the hybridisation of the localisation measurement by means of on-board systems.

(2) *Vehicle data management:* the importance of this function becomes evident in the loading and unloading operations of goods where an exchange certification can be provided by the localisation data on the transport units involved in the

exchange, by the date and time of the exchange, and by the persons concerned who are identified by the transport units for which they are responsible. The certification of dangerous goods exchanges, together with the monitoring of freight conditions, enables the system to prevent the risk of theft or improper use of dangerous goods.

(3) *Communication:* this function enables the exchange of information between the interconnected architectural elements of the system. By applying adequate policies to give priority to communication channels, this function makes it possible to use different data transfer modes, ranging from inter-vehicle to commercial ones which are based on cellular technology, and ensures data exchange between the elements.

(4) *Risk situations management:* when DGT vehicles deviate from their planned route, it is necessary to assess the risk level variation of these new routes. To this end, when planning alternative routes, the system makes sure that the areas concerned do not display high concentrations of DGT vehicles and prevents vehicles transporting incompatible dangerous goods from circulating in the same area at a less than safe distance. In order to guarantee safety in DGT, the system relies on a dynamic risk model which enables it to foresee any risk situations in terms of safety, starting from the current conditions of the monitored system and also covering past risk situations recorded in a dedicated database.

(5) *Monitoring driving behaviour:* safety during DGT is guaranteed by a proper driving behaviour. The system monitors – through satellite survey systems – the trajectory followed by DGT vehicles and in the case of an improper driving behaviour, it sends out warnings to the driver. If the driving behaviour is below the safety limits of the system, it warns safety and emergency management agencies of a possible risk situation. It is evident that such accurate monitoring requires the use of instruments and techniques that can provide vehicle localisation data with a precision level to the order of a few centimetres, which the system is capable of providing. When monitoring driving behaviour, great importance is attached to the monitoring of the driver's condition aimed at assessing whether the driver is in condition to continue the mission. This type of check is carried out by the use of cameras installed in the passenger compartment of the vehicle and by an information analysis model based on ergonomic, psychological and mathematical assessments. The driver must exhibit a driving behaviour suitable to the risk situation in progress or foreseen. For this purpose, the vehicle is outfitted with on-board sensors that monitor the speed of the vehicle, the road conditions, the weather conditions etc., and the driver receives real-time information about the position of other DGT vehicles.

(6) *Minimum-risk planning and routing:* the mission planning function does not target the whole dangerous goods sector, but only authorised DGT which involves the typology of goods classified as most dangerous: explosives, toxic gases, dangerous waste and radioactive substances. For such dangerous goods, therefore, the route that the DGT vehicles will follow must be communicated beforehand to calculate the impact that such carriage will have on the surrounding environment by comparing the planned route with the data stored in a historical risk model. Closely related to mission planning is the idea of route planning, both during the mission design phase and during the journey. This implies the problem of making sure that a given DGT vehicle follows the prescribed route. If the vehicle deviates from the prescribed route, the system must contact the vehicle driver, and if the anomaly persists, it must immediately inform the police of a possible danger situation.

(7) *Position certification:* this feature is particularly important, since the system interfaces with police authorities, emergency managers, and insurance companies. The certification service aims at guaranteeing the validity of the On-board Terminal (OT) position with reference to mobile nodes, providing the control centre with data concerning the accuracy of the certification relative to the OT position. The certification data transmitted by the On-board Terminal ensures that the correct position is exchanged between the OT and the control centre to prevent any accidental or criminal corruption of the information processed by the system. This is made possible by the integrity of the certification chain which is secured firstly by an authentification code created by certification pseudolites to guarantee the pseudolites' signal authenticity, secondly by a tamper-resistant terminal to prevent tampering, and thirdly by a secure communication channel.

(8) *Coordination with the emergency management centres:* the project also deals with the innovative management of emergency situations, including the development of a communications system that provides adequate information for safety and emergency management agencies in order to perform effective and efficient emergency operations.

(9) *Documentation management:* one of the functions of the system is the automatic management of the DGT documentation which works through messaging and notification systems leading to acknowledged and certified operations. This function meets the need voiced by DGT operators to simplify and streamline travel documentation and related procedures, and aims at minimising the paperwork and adding to the security of the travel information. The electronic management of the communication allows DGT operators to cut management costs and receive detailed, real-time information on the state of deliveries, while enabling the system to monitor the progress of the carriage.

The project creates benefits for a number of bodies and institutions:

- For operational and institutional bodies
  - Increased control and statistical governance of the overall DG transportation system
  - Diminished risk of accidents involving dangerous goods (preventive action)
  - Management of emergencies connected with accidents
- For the Italian Space Agency
  - Innovative technology as a prestige element
  - Immediate knowledge and proficiency of new satellite technology
  - Technological innovation in the domain of satellite navigation and its combined use with other technologies (interoperability)
- For the industrial sector
  - Increased competitiveness in the satellite navigation sector
  - Increased skills in the organisational model for the management of DG transportation
  - Integration of Galileo services and components with the ground platform.

6.3.4.2. Rail transport

Location is a very important function for railway applications as well. In order to optimise transport processes, railway operators are interested in fitting their trains and wagons with appropriate location modules for different commercial applications like fleet management, customer information, and punctuality control.

In order to reduce track-side costs as well as to increase safety and line capacity, network operators and railway operators together with suppliers are currently working on new radio-based train control systems which will replace conventional train control and supervision systems. These systems will offer reliable position information not only for control and supervision, but also for non-safety relevant train-borne and track-side applications.

These new systems like the European Train Control System (ETCS) as well as national counterparts like the German FFB (radio-based train control) are designed for high-speed lines as well as for regional lines with high or moderate traffic density. They use a location-based odometry unit as well as balises. The position information is balise-based and has to be transformed to a track-based position (track number and kilometre) for most commercial applications. The use of satellite positioning was not considered for the systems of the first generation foreseen to go into operation from 2004, but now it is definite that control and supervision systems should take full advantage of the possibilities of Galileo.

Furthermore, for regional/secondary lines with low traffic or pure freight traffic as well as for lines where presently no advanced control or supervision system is installed (e.g. most lines in Eastern Europe), concepts for low-cost train control systems are being studied and developed. Because of the reduced performance requirements (which have to be derived), the use of satellite-based locations for these systems seems to be very promising and at present, a lot of tests are being conducted in different countries.

In order to rationalise rail user requirements and thereby define service levels, it was decided to classify the users by a number of different categories:

(a) Safety-critical applications: for this category of user, radio navigation services enable operations that have safety-of-life implications. The capability of the radio navigation service to meet operational requirements is of primary concern to this class of user.
(b) Operationally-critical applications: for this category of user, radio navigation enables operations not directly associated with safety-of-life but linked to other concerns such as economic issues. Services are governed by commercial contracts rather than legislation.
(c) Non-essential use: this category of user has arisen primarily from the opportunistic use of existing systems that are mostly free of charge. Requirements are uncoordinated and services are used on the basis of availability, although some value-added services may be subject to commercial terms of agreement.

A classification system was also defined for the service level requirements in terms of geographical coverage, as follows:

- Global: the service is pervasive to support applications such as long-distance multi-modal fleet management
- Regional: the service level requirement is restricted to particular regions and support applications within those regions, such as inter-urban public transport
- Local: the service level requirement is restricted to particular areas within a region, such as automated tolling.

6.3.5. Concluding remarks

There is an increasing demand for mobility and info-mobility. Space systems, i. e. satellite navigation and positioning, satellite telecommunications and Earth observation from space can help satisfy this rising demand by contributing to safety and security. The examples presented in this paper in three transport sectors (civil aviation, maritime navigation, road and rail transport) demonstrate the help provided by space systems in sustaining mobility, safety and security.

As illustrated, the integration of different technologies can result in a lot of synergies in many application areas. As a result, space applications give appropriate answers to the challenges associated with sustaining and enhancing mobility.

6.4 Telematic solutions based on space technology

Kashif Din, Martin Linauer, Jürgen Zajicek & Martin Reinthaler

6.4.1. Introduction

Space technology is used in many different areas of our life today. A number of applications in the security, environmental, energy and mobility sectors additionally require communication facilities with satellite-based measurement or positioning. In the field of mobility, the positioning of vehicles (or goods) and their tracking and tracing has enabled the development of applications to make traffic safer, cleaner and more efficient. This paper describes two applications that were successfully implemented at Arsenal Research's business unit of Transport Technologies:

- Application 1: optimal route guidance based on Floating Car Data
- Application 2: intelligent route adaptation of hazardous goods.

6.4.2. Application 1: optimal route guidance based on Floating Car Data

6.4.2.1. Overview

Estimations of the actual travel time and potential traffic jams are critical for precise and fast route guidance. One efficient way to measure the traffic status is to use data from vehicles sending their satellite-based position, so-called Floating Car Data (FCD). In the project FLEET (Fleet Logistics Service Enhancement with Egnos & Galileo Satellite Technology), a travel time information service based on FCD for the Vienna region was demonstrated. In this contribution, we will introduce a novel approach to estimating travel times and a multi-constraint route guidance method. The usefulness of the system was illustrated on a test route in the city of Vienna compared with values originating from induction loops.

Traditionally, traffic loads are measured with the help of induction loops. In the past decades, loops have been installed in many cities on the primary network at intersections with traffic light signal controllers. The collected measuring point data are the basic input of a coordinated control system. This system can provide online information on the traffic situation to optimise the trip matrix of a traffic model. The main problem with these loops is that their installation is expensive and that they are sensitive and often subject to damage. Replacing the damaged equipment is costly because the whole road surface must be renewed. Traffic cameras offer an alternative solution to substituting the induction loops. The installation of a camera is comparatively cheap and sometimes provides more reliable data than induction loops. It is possible to detect more lanes (or both directions) with just one camera. Further advantages of cameras are their longer life cycles and the additionally delivered visual information. Camera pictures can be transmitted via websites as well as cellular phones (video streaming, MMS service). As regards roads, an increasing number of commercial and private vehicles equipped with GPS (and navigation) devices have been observed in the last years as well.

In Vienna, the project FLEET was carried out using vehicles as testing cars to determine the traffic status and to provide traffic information. This concept is called Floating Car Data and supplies an excellent database on actual traffic conditions. FCD does not depend on the availability of roadside equipment and is therefore especially well-suited to cover roads that are not equipped with a data collection system. On equipped roads, it provides additional information in the form of directly measured travel times. Earlier research has shown that around 30 vehicles per route and hour need to be equipped to provide a good measurement of travel times, assuming there is no roadside detection equipment.[184] Vehicle data (e.g. speed, position and possible destination) are transmitted to a Traffic Information Server where they are processed and traffic information is sent back to the driver.

These data are transformed into valuable traffic information for travellers. Road users need to reach their destination as fast as possible, but they are also interested in convenient driving conditions represented by traffic density and static parameters (like road type or road length). Static parameters can be obtained from a Geographic Information System (GIS), while traffic density can be measured by induction loops. The most difficult task is to estimate the travel times (or average speeds), which is the main subject of this paper.

First, we generated an efficient road network graph model for the easy handling of special road types (e.g. one-ways) and turning rules (lane connectivity). Afterwards, road element travel times were estimated on the basis of FCD messages by analysing several factors like the speed of the vehicles, the time and number of messages, the speed variance, or the length and type of the road elements.

6.4.2.2. Technical description

In order to optimise trips on road networks, more and more fleet management centres are using systems to track the movement of vehicles, thus allowing individual navigation. Provided that a high overall mileage of a vehicle fleet is tracked, the information generated can also be employed to calculate travel times on a motorway or road network.[185]

6.4.2.3. Floating Car Data provider

The FLEET project demonstrated the functioning of a travel time information service based on Floating Car Data for the Vienna region. This project was co-financed by the Austrian Ministry for Transport, Innovation and Technology under the ARTIST programme for satellite navigation services.

The value-added services developed are based on management data from a Viennese taxi fleet running 800 taxis, all of them equipped with GPS receivers sending their position every 30–45 s. Based on the methods developed and adapted by an information provider (Arsenal Research, Transport Technologies), travel time data for all major streets were calculated. This real-time data on

Fig. 5. *Traffic situation picture of Vienna.*

the present traffic situation was complemented by typical historical time-series of travel times from the pilot trial. In order to validate the reliability of the travel time information, a traffic situation picture (see Fig. 5) was compared to current traffic messages by a radio broadcasting organisation. Both the historical and real-time data were tested for a prognosis of travel times on the major routes in Vienna.

6.4.2.4. Travel time estimation from Floating Car Data

In order to provide drivers with optimal route guidance, the current travel time (or velocity) on each road had to be evaluated. To this end, a method was developed that calculates speed using FCD information on the concerned road, information on the road elements close to the investigated one, and historical data. Road element travel times are estimated on the basis of FCD messages analysing several factors, including not only the speed of the vehicles or the time of the messages, but also information on the road elements, the number of messages, the speed variance and the length and type of the road elements.

Earlier, we summarised several factors of influence regarding optimal speed estimation, including:[186]

(1) Time: data become outdated, i.e., new messages should be given more weight whereas older messages are less significant. However, above a certain threshold of elapsed time, data lose their original coherence with the current traffic state and thus become historical. Historical data can be treated uniformly, with only the type of day and time remaining important.
(2) Message density: the precision of the statement is increased by the number of samples belonging to the road element, i.e. more messages yield a higher weight.
(3) Speed variance: the variance of speed depends on the traffic state. A high variance can be observed for the free flow of traffic, whereas variance is low for congested traffic. Speed values underlie a normal distribution – outliers can be detected on the basis of the properties of the normal distribution (if a sufficient quantity of data is available). Values can be excluded, e.g. if they go beyond two times the standard deviation of speed. All other single values should be considered as normal for the current traffic state.
(4) Type of road element: taking the mean speed for a longer section could lead to imprecise speed information on the individual road elements. This is particularly the case when the estimated path contains different types of roads, with both low and high normal speeds. In this case, the speed of the road with low normal speed could be overestimated and analogously, the speed on a highway could be underestimated by the average speed.

(5) Additional factors: in several cases, estimations for some road elements were insufficient because of a lack of data. However, a travel time estimation should be given for them as well. It can be set to "infinity", which means that these roads are omitted. This approach could lead to very long routes or even to the case that no routes exist in the system. That is why we proposed that the data be estimated on the basis of historical data and travel times estimated by macroscopic models.

6.4.2.5. Evaluation of travel time data

Travel time data from the FLEET project were evaluated on six arterial routes in the city of Vienna with a road length ranging from ~1 to ~18 km. In the inner city, a very good temporal and spatial coverage was observed (more than two vehicles per route and hour, more than 50% of spatial coverage of the route by Floating Car Data). Therefore, a route of the inner city was chosen for a time-series analysis. On other major routes within Vienna, a temporal coverage of 1–2 vehicles per route and hour and a spatial coverage of between 30% and 50% of the route were achieved, which is a good basis for measuring route average speed. The temporal and spatial coverage of city motorways yielded less than one vehicle per route and hour and partly less than 20% of spatial coverage of the route, which only permits an estimation of the current speed on the route.

The trend of the average route speed was compared to the local density of an induction loop located on the same route. It was observed that the average route speed decreased with an increase in local density, and vice versa. Observations

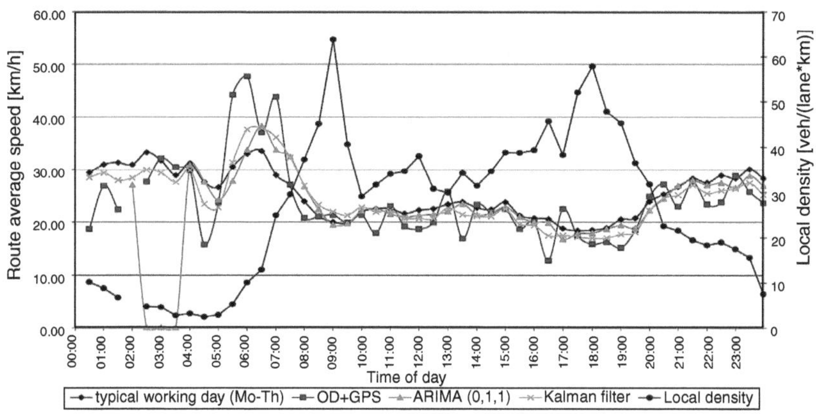

Fig. 6. *Average speed on a typical working day, measured by our proposed method considering both OD and GPS values, using an ARIMA*[190] *model and a Kalman filter.*

showed a clear reduction of average route speed for approximately 12 h on a typical working day.

Results of a prediction with ARIMA modelling showed an average relative error of the predicted value at $t+1$ vs. the measured value of 10–22%. The error for a prediction with the last measured value ranged between 16% and 28%, and for a prediction with the last observed difference between 15% and 28%. The prognosis error was markedly higher when values from the current time series were missing (see Fig. 6).[187] ARIMA models and their application for traffic data have been described elsewhere.[188,189]

The application of the typical time-series' value yielded an average relative error between 11% and 21%. This underlines the high information content of typical time-series for regularly occurring changes in the speed level. However, it is important to consider current measurement values in order to detect traffic interruptions (as, for example, road works or accidents).

The results of the estimation with Kalman filters showed an average relative error of the corrected value at time $t+1$ vs. the measured value of between 9% and 17%. When only the predicted value at time $t+1$ of the second model was compared to the measured value, the average relative error ranged between 14% and 20%. Thus, correction with the current values of the average route speed improved the prediction by 4–5%. As for ARIMA models, the prognosis error increased by 3–7% when values from the current time-series were missing.

6.4.2.6. Results

We have proposed a novel adaptive route choice algorithm based on travel time estimations from floating cars. Compared to traditional methods, this improves the traffic information for each driver and therefore the information basis for individual decisions, especially in the case of traffic jams. Judging from our experiments, some hundreds of floating cars (vehicles of taxi fleet) are enough to calculate a good picture of the traffic situation in a city. However, with an increase in data suppliers, the picture will be more precise.

6.4.3. Application 2: monitoring of hazardous goods with real-time route analysis

6.4.3.1. Overview

The fleet management systems used by many companies are essential components of their daily business. Using Floating Car Data (FCD) technology for travel time calculations and fleet management could open even more and new possibilities in the field of Intelligent Transport Systems (ITS).

On the Austrian roadways, approximately 20 accidents with hazardous goods transports occur every year,[191] posing a potentially high threat to people and the environment. The system developed by us and presented here focuses on this kind of transport and calculates detailed routes and travel times for each hazardous goods transport over the road network. The expert system also recognises meeting points with other high risk transports and broadcasts warnings to the relevant drivers. In addition, the FCD is used to calculate actual travel times and historical time-series. The project included simulations and a field trial done in the city of Vienna.

The technology employed in order to make use of mobile traffic participants as sensors is referred to FCD. FCD are currently used to calculate either historical time-series on specific road sections, or current travel times, or for surveying origin-destination matrices. The analyses of driving behaviour, the generation of graph-based maps,[192] as well as the integration of data in traffic models and traffic management centres are further areas of application.

In the ongoing development from FCD to extended FCD (xFCD), positioning data are complemented with additional attributes. The measurement values of vehicle-based sensors can also offer additional information, for instance about the driving behaviour of a specific driver.

The quality and availability of data are essential criteria in the devising of fleet management systems. They are on the one hand defined by the type and dimension of the fleet and by the technology employed for determining their position, and on the other hand by the communication method used by the system. The present system is operated with a fleet management system which offers xFCD in real-time. Information about the kind of vehicle, the cargo, and the origin and destination are available in the system. This allows for both analysing the trajectories of traffic participants and for recognising traffic states on all single sections of the track.

6.4.3.2. Technical description

Several existing technologies for the positioning of objects and for data communication are available. The primary objective of the system developed is to make use of the existing technologies and exploit their information content by using specific algorithms. The architecture of the system uses proven positioning devices and communication technologies and includes flexible interfaces.

6.4.3.3. System architecture

The system architecture is divided into analysis and the calculation of current traffic states. While route analysis is based on current trips, the built-in Floating Car Evaluator (FC evaluator) analyses already stored and corrected FCD trajectories. Both modules are complementary as far as the analysis of routes is based on traffic states coming from the FC evaluator, which are calculated based on the obtained trajectories of the route analysis.

The map-matching algorithm establishes a relation between the positioning data and the digital map. The indication of two-dimensional coordinates is a minimum requirement. Additional information like the heading or quality indications, which are often only available in satellite-based systems, is able to substantially improve the results of the procedure.

Basically, route analysis can be performed with historical FCD as well as with current trips. The system developed aims for an anticipatory route analysis that requires trips with a known destination in order to be able to calculate the chosen route already at the starting point of the trip. Essential information that must be known is the kind of fleet and vehicles (hazardous goods transports (HGTs), passenger cars, etc.) to be able to calculate the route as exactly as possible. Infrastructure-related parameters, like the Functional Road Class (FRC), diverse driving bans for specific vehicle classes, speed limits, turning relations, velocities, and even more decisive information is required for the routing module. The routing module uses an adapted Dijkstra algorithm which is also called Radix Heap implementation and presents a mixed form of the original algorithm and the Dial implementation.[193]

Additional information can be gained from historical trips that can act as an input to the choice of routes. Not only the kind of vehicle (passenger car or a HGT), but also the economic background of the fleet will influence the driving behaviour. Thus, the trips of taxis (drivers are familiar with the surroundings) will strongly differ from the trips of action forces (blue lights) or from drivers who use navigation systems. The best method is the evaluation of the respective fleet (number of datasets, 85th percentile of velocities, etc.) in order to discern the driving behaviour and to set weighing parameters for the map graph properties (nodes and edges).

The module scheduler of the expert system establishes a time relation with the route. The calculated schedules can be based on different kinds of velocity values – depending on their availability, legally prescribed velocities, current FCD-based velocities, free flow velocities calculated on the basis of historical data, and velocities according to the FRC class (lowest accuracy).

In order to improve the accuracy of the expected travel times, current or historical values obtained with the Floating Car Evaluator based on the trips already performed by the entire fleet should be used. The Floating Car Evaluator calculates mean velocities for the covered trajectories and aggregates all existing values per road section and per time interval.[194] Data from several fleets can be introduced as a basis for the FC Evaluator to improve the coverage rate (number of cars in relation to overall traffic). In order to generate historical velocity time series, FCD have to be available over long time periods. However, the time series are useful for filling the gaps of the current velocity calculations.[195] In the scheduler, simple prognosis methods were also implemented for short prognosis horizons in order to estimate travel times at the expected time of the trip.

The results of the routing algorithm and of the scheduler are combined into vehicle trajectories. The trajectories are attributed to the map and marked with time periods for the expected time of the trip. Based on current trajectories, different analyses can be performed, like the detection of meeting points of FC of the same category or the recognition of deviations from the route or from the time schedule.

6.4.3.4. Use cases and field trial (project SHAFT)

The expert system developed was used in a road safety project and evaluated by simulation and field trials. In this context, the project SHAFT (Standardized Hazardous Goods Transport Alerting Field Trail) focused on the development and demonstration of a monitoring system for heavy and hazardous goods transports (HGTs) which includes a tracking and tracing system controlling all HGTs on the Austrian road network over the whole transport chain.

SHAFT introduced a standardised electronic waybill. The electronic management of waybills improves the monitoring capabilities of operators and carriers, allows for the use of transport data in intelligent transport systems, and creates economic benefits. These are the reasons why electronic waybills play a central role in SHAFT.

The entire system is based on the information of each HGT kept in waybills stored in a Fleet Management System. The interface to the On-Board Units (OBUs) delivers the position data to the server and allows the driver to read and edit the waybills. For the shipper or the transport company, a Web client was implemented to control the fleet and create the waybills. Essential information, like the origin and destination of the transport and the cargo of the hazardous goods vehicle, are available immediately at the start of the trip.

One purpose of SHAFT was to provide important information to action forces especially in the case of accidents. The knowledge about the amount and type of the transported hazardous goods and the position of an accident can considerably improve the efforts of the fire brigade. SHAFT thus includes an interface to a hazardous materials matrix, which contains the risk potentials of all hazardous goods categorised by chemicals. The system assesses the risk potential of each transport based on the information of the cargo and allows decisions to be taken in the case of detected meeting points of two or more HGTs travelling on the same sensitive road section. The SHAFT system stops the transports with lower risk levels and gives priority to the "high risk" HGT to pass through a sensitive section. This keeps the potential impact on the environment low.[196]

The use of the routing algorithm and the scheduler included in the expert system offers detailed information on the vehicle trajectories. The system architecture includes the most efficient handling of the system resources in order to keep calculation times low and be able to monitor a large numbers of vehicles. To keep calculation times low, it is essential to reproduce the driving behaviour of hazardous goods vehicles as precisely as possible.

All hazardous goods vehicles are equipped with the required hardware devices. Positioning data are broadcasted by the OBUs in flexible time intervals (currently 60 s) or after passing a minimum distance of 100 m since the last broadcasted position (congestion, waiting time at toll stations, etc.). The position information is used to evaluate and correct the trajectories. Ideally, the trajectory obtained is confirmed by the broadcasted positioning data and no correction is required. Only

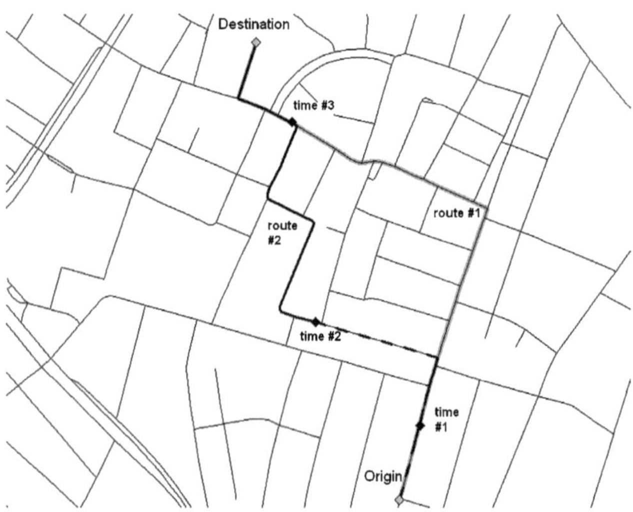

Fig. 7. *Route adaptation of a HGT during its journey.*

when the actual GPS-Position of the vehicle is not referenced to the calculated route is the routing module required.

Figure 7 shows the route adaptation of a HGT during its journey. At time #1, route #1 was calculated by the expert system for the journey to the destination. At time #2, the OBU sent a position which differs from the proposed route. Based on the information, the expert system calculated the new route #2. At time #3, the position corresponded to the proposed route #2 and the system confirmed the current route to the specific transport.

The system analyses the trajectories of all HGTs on the relevant road network and makes decisions on how to handle meeting situations of HGTs at critical sections on the road network. As mentioned above, the system will stop HGTs with a lower risk potential to give priority to the "higher risk" transports.

The basic reference tool is a road map with geographically referenced risk areas, such as single-tube tunnels, water protection areas, etc., which are used for the analysis of the route. Meeting points in nature or water protection areas are also classified as high-risk. Moreover, parking spaces and rest stations are indicated to allow drivers to stop due to a detected meeting point with a high-risk HGT.

The meeting area calculated by the expert system is shown on the map in Fig. 8. The vehicle with OBU ID1 is driving from the upper left to the lower right, indicated by squares for the GPS positions and a dotted line for the calculated trajectory. The second vehicle is indicated by circles and the anticipatory route is printed (grey line).

The system analyses the two route trajectories and finds a geographical and schedule meeting point on road id 9688 (black line). This meeting point is handled as a high-risk meeting point because the road section is marked as a tunnel. The

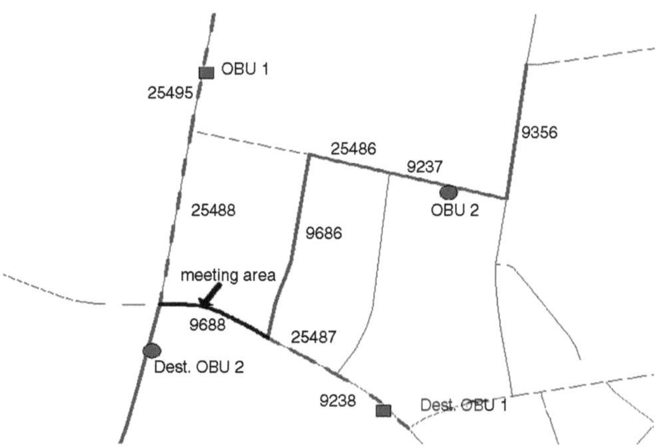

Fig. 8. *Calculation of a meeting area by the expert system.*

experts system determines the risk potentials and sends warnings to the On-Board Units.

6.4.3.5. Results

The system developed can help to avoid accidents of hazardous goods transports or at least provide exact information to action forces. The use of an electronic waybill helps to handle transport chains more efficiently and use the obtained information in telematics systems. Furthermore, the tests of the system show that the data produced by fleet management systems can produce much more than only information on the current position of the company's vehicle. The field trial in Vienna was very successful and confirmed the functioning of the system that met all expectations.

6.4.4. Conclusions

With the applications above, two very promising demonstrations of the use of space-based positioning data were demonstrated. Positioning with space technology is essential for the development of highly sophisticated mobility applications. The high market penetration of aftermarket navigation products enhances the development of end user services and increases the number of vehicles sending Floating Car Data. Since security and safety are constant requirement themes, a reduction of incidents and dangerous situations can be achieved by warning drivers or by avoiding the meeting of traffic participants with high risks in sensitive areas (tunnels, water-protection areas, residential areas, etc.). It may be expected that with the higher market penetration by navigation products, this positive development in terms of safety will continue as well. The introduction of Galileo will improve the positioning and availability of signals and this will definitely lead to highly sophisticated applications that make traffic safer, more efficient, and more environmentally-friendly.

[184] Davidsson, Fredrik, Matstoms, Pontus, et al. OPTIS Statistical Analysis and PROBE Simulation Study. Stockholm: Royal Institute of Technology, 2002.
[185] Ibid.
[186] Laborczi, Peter and Linauer, Martin. "Optimal Route Guidance Based on Floating Car Data." Paper presented at the 10th World Congress on Transport Research, Istanbul, 2004.
[187] A Kalman filter is a combination of equations for estimating the state of a time discrete process. The process is described by an autoregressive term, a control value and a random value (process variation).
[188] Box, George E.P., Jenkins, Gwilym M. and Reinsel, Gregory C. Time Series Analysis: Forecasting and Control. Englewood Cliffs, N.J.: Prentice Hall, 1994.

[189] Schneider, Michael. Mathematische Modelle zur Prognose der Verkehrslage. Diploma thesis, TU Wien, 2003.
[190] One approach of stochastic models are the ARIMA (Auto Regressive Integrated Moving Average Models). They consist of an autoregressive and a moving average term.
[191] Dietl, Eva and Esberger, Robert. Straßenverkehrsunfälle 2004. Vienna: Statistik Austria, 2005.
[192] Laborczi, Peter and Zajicek, Jürgen. "How to Generate a Graph-Based Map from Floating Car Data." CD-ROM Proceedings of the 10th World Congress and Exhibition on Intelligent Transport Systems and Services, Madrid, Spain, 16–20 Nov. 2003.
[193] Ahuya, Ravindra K., Magnangi, Thomas L. and Orlin, James B. Network Flows: Theory, Algorithms, and Applications. Englewood Cliffs, N.J.: Prentice Hall, 1993.
[194] Laborczi, Peter. Optimal Route Guidance Based on Floating Car Data. Vienna: Arsenal Research, 2004.
[195] Ibid.
[196] Zajicek, Jürgen. "Developing a Hazardous Goods Monitoring System for the Austrian TERN – Project SHAFT." Paper presented at the 31st FISITA World Automotive Congress, Yokohama, Japan, 22–27 Oct. 2006.

CHAPTER 7

SECURITY

7.1 Overview

The context: traditional peace-keeping and peace enforcement are no longer sufficient in today's world. The evolution of threats has led to the emergence of the new concept of "sustainable security", which is now needed. It includes protection, reconstruction, CIMIC (Civil and Military Cooperation), disarmament and state-building. Furthermore, security is needed on Earth, but also in outer space.

The contribution of space: space systems provide unique elements to ensure security on Earth. For instance, space-based applications offer precise maps for peace-keeping, military and intelligence purposes, as well as means for checking compliance with international treaties. Space plays a role in ensuring security in a broader sense, as it contributes to food security or environmental security. At the same time, security in outer space is needed: the space environment has to be preserved and the use of force in outer space has to be discouraged. Given the unique contribution of space systems to security on Earth, ensuring security in outer space is also a prerequisite for security on Earth.

The papers: the three papers of this section provide two different but complementary approaches to space, security and sustainability. First, the need for sustainable security is explained, and subsequently, both the contribution of space systems to security on Earth and the need for security in space are reflected upon.

In "The European Contribution to Sustainable Security", Heinz Gärtner from the Austrian Institute for International Affairs reflects on the definition of the concept of security and identifies the new threats to security. He underlines the emergence of an entirely new concept of security, i.e. the concept of sustainable security, and summarises the conditions necessary for such security to be achieved. He then reviews the recent evolution and future prospects of the concept of European security by analysing key elements of the European Security and Defence Policy (ESDP) and the possible evolution of the relations between Europe and NATO. The paper concludes with the new approaches needed to achieve sustainable security.

In "The Global and European Security Context: A Proposal for Space", Erwin Duhamel from ESA describes the actual and potential contribution of space assets to security policy. He first reviews the global and European security context and then assesses a possible future for the European use of space for security. He also details the current use of space systems to support Europe's security and finally focuses on the aim and scope of a European space doctrine.

In "Space for Security and Security in Space", Gérard Brachet, Space Policy Consultant and Chairman of the United Nations Committee on the Peaceful Uses of Outer Space (UNCOPUOS) first reviews the current international legal framework for space activities, the recently adopted "Space Debris Mitigation Guidelines", the "Rules of Good Behaviour" for space activities under preparation, and the issue of space weaponisation. He then examines the various contributions of space systems to security on Earth. His paper concludes with the need for "space security", i.e. the ability to use space in a sustainable manner, and for "best practices" adopted at the international level in order to ensure that the use of space for security is not jeopardised by the lack of security in outer space.

7.2 The European contribution to sustainable security

Heinz Gärtner

7.2.1. Security and philosophy

What is security? Security exists if there is either no threat or if there are sufficient capacities to defend oneself against any threat. It can be achieved by increasing the capacities or by abating the threats. Hence, traditional Cold War concepts of security emphasised that lowering one's probability of defeat through increasing one's military power led to more security. Today, 'threat-abatement' rather than 'capacity enhancement' meets the requirement of the post-Cold War security environment. It is based on a comprehensive concept of security that ranges from military to humanitarian actions.

In philosophical discourse, there are two opposing ideas: in a Hobbesian world, states are enemies. The world is anarchic without any regulating authority. States build capacities. Security means waging wars or – if successful at all – building a system of a balance of power.

By contrast, in the Kantian world, states are friends. Rather than building up military capacities, states abate threats as a means of crisis management and conflict prevention. It is not a balance of power that generates stability, but a federation of states. In the 21st century, the real danger is no longer the anarchy of the Hobbesian world. It is not an all-out war in the world of states that constitutes a threat, but rather the lack of legitimate governance in many states. Security is about building functioning states.

7.2.2. The new threats

Javier Solana, the High Representative for Foreign Policy of the European Union, presented a European Security Strategy (ESS)[197] that was adopted by the European Council in December 2003. The document asserts a need for a role for the European Union in global security and in any actions, including military force, deemed necessary to achieve that security. It calls for member nations to increase their military abilities and share their military resources. The document defines the proliferation of weapons of mass destruction, terrorism, failed states, and uncontrolled immigration as the principal new challenges. First, there are dysfunctional (e.g. failed or failing) states which often, but not always emerge from post-conflict societies.

These states are:

(a) The source of humanitarian disaster,
(b) Breeding grounds for organised crime and terrorism,
(c) Areas for illicit weapons and drug trafficking, and even sources of weapons of mass destructions (WMD),
(d) Lost areas for good investments and trade, and
(e) They produce large refugee flows which subsequently lead to uncontrolled immigration.

7.2.3. Sustainable security

Sustainable security requires as a minimum the creation of a stable society with good governance and a working administration. The goal should be functioning states with rule of law, civil administration, democracy, and the legitimate monopoly of the use of force (Max Weber).

How can sustainable security be achieved? In the 1990s, it became clear that traditional peacekeeping with the consent of the conflicting parties is not sufficient. What is needed in pre- and post-conflict societies is peacebuilding, reconstruction and disarmament. Demobilisation, reintegration (DDR) and security sector reform (SSR) are necessary, which involve both military and civilian means. What are these means?

(1) It starts with disaster relief (as in the cases of the Asian tsunami of December 2004, hurricane Katrina, or earthquakes). New areas of cooperation emerge. Civilian and military personnel have to train and work together in civil-military coordination and cooperation (CIMIC). The EU has developed a civil protection mechanism. People have to be flown out of crisis areas, and food, water, medicine and shelter must be brought in.
(2) States have the "responsibility to protect" their citizens in cases of genocide, massive human right violations, famine, poverty, and in the long run from the consequences of climate change. The primary responsibility lies with the state itself. If the state is unwilling or unable to do so, it yields to the international community and international organisations (e.g. UN, OSCE).

(3) To build sustainable security, there has to be a functioning government endowed with the legitimate monopoly of the use of force. It has to be based on the rule of law and good governance, with a working administration and ideally with democratic legitimacy. These are the core pillars of state-building.

7.2.4. The European Security and Defence Policy (ESDP) in the Lisbon Treaty

The EU Heads of State failed to adopt the new European Constitution in 2005, but did agree on the European Security and Defence Policy (ESDP) in the Lisbon Treaty with some adjustments. This document recognises that the concept of security is very broad and by nature indivisible. It is a concept of security that goes beyond purely military aspects and covers not only the security of states, but also the security of citizens. It gives the Union military options over and above the civil instruments of crisis prevention and management. The text of the Lisbon Treaty will not be much different from the text of the Constitution.

The ESDP will continue to work with intergovernmental structures ("In particular, national security remains the sole responsibility of each Member State" [3a])[198] and without a qualified majority. Decisions will continue to be taken unanimously:

> "Decisions relating to the common security and defence policy, including those initiating a mission as referred to in this Article, shall be adopted by the Council acting unanimously on a proposal from the High Representative of the Union for Foreign Affairs and Security Policy or an initiative from a Member State." (28A, 3)

The opting out by a member state does not prevent the EU from taking action, but rather means that ESDP activities will be more effective. The ESDP will develop specific provisions for implementing the ESPD, including future steps towards the framing of a common Union defence policy. These steps could lead to a common defence when the European Council, acting unanimously, decides it is in accordance with the member states' respective constitutional requirements.

The Treaty underlines that it respects the essential state functions, "including ensuring the territorial integrity of the State, maintaining law and order and safeguarding national security. In particular, national security remains the sole responsibility of each Member State" (Art. 3a).

The following cornerstones will remain explicitly or implicitly enshrined in the Treaty:

(1) The institutional structures (Political and Security Committee, Military Committee, Military Staff) are already in place.

(2) The tasks of the Union, which are the extended Petersberg missions, include joint disarmament operations, humanitarian and rescue tasks, military advice and assistance tasks, conflict prevention and peacekeeping tasks, tasks of combat forces in crisis management including peacemaking, and post-conflict stabilisation. All of these tasks may also contribute to the fight against terrorism, including support to third countries (e.g. the U.S.) in combating terrorism in their territories:[199]

> "The common security and defence policy... shall provide the Union with an operational capacity drawing on civilian and military assets. The Union may use them on missions outside the Union for peacekeeping, conflict prevention and strengthening international security in accordance with the principles of the United Nations Charter. The performance of these tasks shall be undertaken using capabilities provided by the Member States." (28A, 1)

(3) The Lisbon Treaty envisages that those member states whose military capabilities

> "fulfil higher criteria and which have made more binding commitments to one another in this area with a view to the most demanding missions shall establish permanent structured cooperation within the Union framework." (28A, 6)

Member states will harmonise the identification of their military needs, by pooling and, where appropriate, specialising their defence means and capabilities, as well as by encouraging cooperation in the fields of training and logistics. They will also take concrete measures to enhance the availability, interoperability, flexibility, and deployability of their forces, in particular by identifying common objectives regarding the commitment of forces. This may include reviewing their national decision-making procedures. The Union should be provided with an operational capacity that draws on civil and military assets to use such assets "outside the Union for peacekeeping, conflict prevention, and strengthening international security in accordance with the principles of the United Nations Charter".[200]

The two exceptions included are that the common security and defence policy of the Union should not prejudice the specific character of the security and defence policy of specific member states, and that the common security and defence policy of the Union respects the obligations under the North Atlantic Treaty of those member states which perceive their common defence as realised in the North Atlantic Treaty Organisation.

The battle group concept might well fit into this structured cooperation. In line with a British–French proposal that member states should create rapid deployment forces or battle groups by 2007 that could be sent on emergency missions on behalf of the EU, notably to Africa, the EU should have the capacity to supply targeted combat units for the missions planned. These units should be structured at a tactical level as combat formations, with support elements that include transport and logistics capable of carrying out the tasks referred to above (extended Petersberg Tasks). These capabilities could operate at either the national level or as a component of multinational force groups, in particular, in response to requests from the United Nations Organization under Chapters VI and VII of the UN Charter (though not exclusively and necessarily).

Such battle groups would be composed of up to 1,500 troops ready to be deployed for combat within a period of 5–30 days (e.g. in jungle, desert and mountain operations), "in particular in response to requests from the United Nations Organizations" and could be sustained for an initial period of 30 days, with possible extensions to up to at least 120 days. It might very well be that efforts to achieve the Headline Goal Force will slow down under the new provisions, because military capabilities will not extend to the soldiers and equipment needed within this framework.

(4) A European Defence Agency (EDA) to assist the Council in evaluating the improvement of military capabilities has already been set up based on the Treaty of Nice and has started to work:

> "Member States shall undertake progressively to improve their military capabilities. The Agency in the field of defence capabilities development, research, acquisition and armaments . . . shall identify operational requirements, shall promote measures to satisfy those requirements, shall contribute to identifying and, where appropriate, implementing any measure needed to strengthen the industrial and technological base of the defence sector, shall participate in defining a European capabilities and armaments policy, and shall assist the Council in evaluating the improvement of military capabilities."

(5) There is a clause on security commitments in the Treaty:

> "If a Member State is the victim of armed aggression on its territory, the other Member States shall have towards it an obligation of aid

and assistance by all the means in their power, in accordance with the self-defence provisions of Article 51 of the United Nations Charter." (28A, 7)

Before this clause was enshrined in the failed European Constitution, the neutral and non-aligned states, Finland, Sweden, Ireland and Austria expressed reservations with regard to this clause which establishes mutual defence in the event a member state is attacked. These countries wanted to preserve their status. Therefore, they were successful in achieving the inclusion of the provision that "this shall not prejudice the specific character of the security and defence policy of certain Member States".

The NATO member states on their part were afraid that these new commitments could interfere with their obligations under Art. 5 of the Washington Treaty and claimed exemption for themselves:

> "Commitments and cooperation in this area shall be consistent with commitments under the North Atlantic Treaty Organization, which, for those States which are members of it, remains the foundation of their collective defence and the forum for its implementation." (28A, 7)

(6) A proposed solidarity clause is designed to help prevent terrorist threats. Should a Member State fall victim to a terrorist attack or a natural or man-made disaster, the other Member States shall assist it at the request of its political authorities:[201]

> "The Union and its Member States shall act jointly in a spirit of solidarity if a Member State is the object of a terrorist attack or the victim of a natural or man-made disaster." (188R)

It is important to note that this clause is not part of the ESDP. Therefore, it should not be confused with collective defence commitments (28B, 7). The Member States shall provide assistance only "at the request" of a Member State and its political authorities", which is different to the alliance obligations of collective defence.

7.2.5. NATO's transformation

The future of European security has to address the EU–NATO relationship. There is a constant debate on whether NATO and ESDP are competitive, complementary or duplicating. First, we have to look at NATO itself, however. Based on the assumption that alliances cannot survive without a clear and unifying threat, some analysts predicted NATO's imminent demise after the end of the East–West conflict. No alliance in history has survived for long after the disappearance of its enemy. NATO did not expire, but it has had to adapt and transform itself more and more, and along the way, it effectively jettisoned collective defence as the core rationale for its existence.

Since the end of the East–West conflict, NATO has undergone a significant transformation process, one that has been speeded up by the terrorist attacks of 11 September 2001. After 9/11, NATO was again at risk of becoming irrelevant in a world in which terrorism had become the principal strategic threat. Founded as a collective defence organisation at the onset of the Cold War, NATO had to revise its strategic concepts to respond to a broader spectrum of the threats.[202] NATO was going to lose its traditional role, or at least change it radically, as an instrument of collective defence. It has become clear that European governments will support the United States in crisis management operations, even if they take place outside of NATO's core area, such as in the Balkans, Afghanistan, and under certain conditions even in Iraq.

The attacks of 9/11 had also another impact: NATO became an organisation with global reach. Not only was Article 5 transformed after 11 September, but NATO's role definitively became global and made the geographical restrictions of Article 6 of the Washington Treaty meaningless. The final document of the NATO Prague 2002 meeting states that "in order to carry out the full range of its missions, NATO must be able to field forces that can move quickly to wherever they are needed . . . to sustain operations over distance and time". NATO's Secretary General[203] concluded that "the 'out-of-area' debate has been resolved once and for all". In Afghanistan, where NATO leads the international stabilisation force, and in Iraq, where it will probably support the stabilisation of post-Saddam Iraq, NATO not only acts 'out-of-area' but 'out-of-continent.'[204] NATO will not scrap Article 6, in the same way as it will not scrap Article 5.

At the Prague Summit, the heads of governments approved the formation of a NATO Response Force (NRF) of around 20,000 troops to deal with the new threats of the 21st century. A brigade (usually a military unit of about 5,000 troops) should be deployable wherever it is needed within 5–30 days. The new NATO Response Force is intended to be a coherent, high-readiness, joint, multinational force package, technologically advanced, flexible, deployable, interoperable, and sustainable. It will not be a permanent or standing force. Rather, its component units will be reconfigured as required to meet the needs of a specific operation. The NRF will be able to carry out certain missions on its own, or serve as part of a larger force to contribute to the full range of Alliance military operations. Within the framework of the Prague Capabilities Commitment, individual countries will have to commit themselves within pre-set deadlines to providing specific equipment and expertise. The NRF and the European Union's Headline Goal Force are intended to be fully compatible and mutually reinforcing initiatives. U.S. Senator Luger[205] sees NATO's future in its global role:

> "The time when NATO could limit its missions to the defence of continental Europe is far in the past. With the end of the Cold War, the gravest threats to Europe and North America originate from other regions of the world. And this requires Europeans and North Americans to be bolder in remaking our alliances, forging new structures and changing our thinking. To be fully relevant to the security and well-being of the people of its member nations, NATO must think and act globally."

NATO is increasingly becoming involved in disaster relief and humanitarian relief operations. In response to U.S. requests after hurricane Katrina, NATO's Euro-Atlantic Disaster Response Coordination Center (EADRCC) coordinated relief support in the form of food supplies.[206] Following the earthquake in 2005, the North Atlantic Council (NAC) approved an air operation to bring supplies from NATO and partner countries to Pakistan for assistance in the relief effort. This decision followed a formal request for assistance by the Government of Pakistan. NATO nations decided to put into place a strategic air bridge to Pakistan to move urgently required relief assets,[207] and marked the largest operation of the force which has also been used to help protect elections in 2006 in Afghanistan, guard the Athens Olympics, and coordinate an airlift of European aid to the United States after hurricane Katrina.

NATO is training Iraqi and Afghan security forces. At the African Union's request, NATO assisted the African Union (AU) Mission in Sudan (AMIS). It supported the African Union's peacekeeping operation in Darfur with airlifts. It has been expanded at the African Union's request with additional AU appeals to NATO for airlift support to the AU Mission in Somalia and assistance to the African Standby Force.

NATO's Secretary General[208] recognises the changing character of NATO that is different to the "old NATO . . . during the Cold War. . . . Over the past few years, NATO has adapted to the new, 21st century risks and challenges and turned into a very flexible security instrument. . . . While collective defence remains the core purpose of NATO, we all realise that we no longer need to defend Western Europe against the threat of a massive invasion from the East. We all agree that a purely territorial understanding of security is simply too narrow to cope with the new security challenges before us".

A group of former senior NATO officers argue that NATO has not completed the transformation from a Cold War organisation to one capable of dealing with threats and conflicts in the 21st century yet. The officers' 152-page report, "Toward a Grand Strategy for an Uncertain World"[209] calls for a radical overhaul of the alliance, which they say is paralyzed by cumbersome decision-making rules, inequitable funding arrangements and an inability to sustain long-term missions. After explaining the complexity of the threats, the authors assess current capabilities and analyse the deficiencies in existing institutions, concluding that no nation and no institution is capable of dealing with current and future problems on its own. The only way to deal with these threats and challenges is through an integrated and allied strategic approach, which includes both non-military and military capabilities. Based on this, the authors propose a new grand strategy that can be adopted by organisations and nations alike, and then look for options on how to implement such a strategy. They conclude that given the challenges the world faces, this is not the time to start from scratch. Thus, existing institutions, rather than new ones, are our best hope for dealing with current threats. NATO needs more non-military capabilities and this underpins the need for better cooperation with the European Union. The authors offer ideas on how to solve the problem of the rivalry with the EU, and how to give NATO access to instruments other than military ones. In the long term, the authors propose abandoning the two-pillar concept of America and Europe cooperating and suggest aiming for the long-term vision of an alliance of democracies ranging from Finland to Alaska. To begin the process, they propose the establishment of a directorate consisting of the USA, the EU and NATO. Such a directorate should coordinate all cooperation in the common transatlantic sphere of interest.

7.2.6. New approaches for the armed forces

In many areas, armed forces will have to adopt new security approaches. First, military capabilities mean less and less war-fighting alone but increasingly include the organisation and training aimed at conducting low-intensity operations, keeping the peace and assisting in reconstructing societies after wars and violent conflicts, and providing humanitarian action and disaster relief. The task of the armed forces is no longer primarily the defeat of an enemy, but rather the protection of its people. Armies should give up their unwillingness to contribute troops for 'constabulary' duties, peacekeeping, and nation-building.[210] The wars in Iraq and in Afghanistan have shown that a high-intensity war with advanced combat technology is ultimately not sufficient. For post-conflict reconstruction, peacekeeping, state-building and lower intensity capabilities, specially-equipped and trained armed and police forces are necessary – a chastening lesson U.S. troops[211] learned during the occupation of Iraq.[212] The U.S. is already changing its attitude towards post-conflict reconstruction and nation-building. Based on its experiences in Iraq, the U.S. has brought in more mobile and less obtrusive units, staffed by soldiers with some knowledge of the local language and customs based on 'culturally sensitive training' to replace heavy armoured units.[213]

Second, military forces are increasingly involved in disaster relief efforts after natural catastrophes such as during the Indian Ocean tsunami, hurricane Katrina and the earthquake in Pakistan in October 2005. Disaster assistance is about logistics – moving people, water, food, medical supplies and heavy equipment to save lives and communities. The distinctions between war and relief, between domestic and foreign deployments, are breaking down.

This means that an entirely new concept of security is emerging. It is not a world in which the destruction of state enemies is the primary goal as in previous centuries or in the Hobbesian concept where states always have to be prepared to be vanquished or to defeat their enemies. Rather, the opposite of the Hobbesian principle is sustainable security. It is about the protection of people and saving lives. It is about human security that puts the individual at the centre of security and safety. It is about building societies and not destroying them.

---

[197] Solana, Javier. "A Secure Europe in a Better World." Speech at the European Council, Thessaloniki, 20 June 2003.

[198] European Union. "Treaty of Lisbon Amending the Treaty on European Union and the Treaty Establishing the European Community, Signed at Lisbon, 13 December 2007." Official Journal of the European Union C 306, 17 Dec. 2007.

[199] These are the so-called Petersberg Tasks that were proposed by the Western European Union at Petersberg, Germany in 1992. Subsequently, they were incorporated into the Amsterdam Treaty of the European Union. The Draft Constitution extended them to include disarmament efforts, conflict prevention, and terrorism.

[200] Protocol on Permanent Structured Cooperation, established by Article 28A I-40(6) of the Treaty on European Union.

[201] The European Council already anticipated this solidarity clause contained in the Constitution of the Joint Declaration of 25 March 2004. On March 26, the heads of state and government adopted the clause independently of the Constitution. This clause must not be confused with a mutual defence commitment, since it applies to non-state terrorism and is not part of the chapter of the ESDP.

[202] How broad NATO's mission has become is shown by the fact that NATO helped with security during the 2004 Olympics in Greece.

[203] Robertson, George. "Europe's Transformation." Speech at the Conference of the Aspen Institute Berlin and the NATO Host Committee for the Prague Summit, Prague, 20 Nov. 2002.

[204] For the first time ever, five NATO nations also held aerial war games with non-members, India and Japan, under an exercise codenamed Cooperative Copethunder in Alaska in July 2004.

[205] Senator Richard Lugar on the vital role of NATO in Afghanistan at a hearing on 31 Jan. 2008.

[206] NATO. "Statement by the NATO Spokesman on Support to the US in Response to Hurricane Katrina." Press Release 106, 4 Sept. 2005.

[207] NATO. "Statement by the Secretary General on NATO Pakistan Earthquake Relief Operation." Press Release 123, 11 Oct. 2005.

[208] Scheffer, Jaap de Hoop. Speech at the Emirates Center for Strategic Studies and Research, Abu Dhabi, 24 Jan. 2008.
[209] Naumann, Klaus, et al. Towards a Grand Strategy for an Uncertain World: Renewing Transatlantic Partnership. Lunteren: Noaber Foundation, 2007.
[210] See: Dobbins, James. "Next Steps in Iraq and Beyond." Testimony before the Committee on Foreign Relations, United States Senate. 18 Sept. 2003. http://foreign.senate.gov/testimony/2003/DobbinsTestimony030923.pdf. James Dobbins rightly criticises that, as to *nation-building*, "if there is any lesson to be learned from our "post-conflict" involvement in Iraq . . . it is that we have failed to adequately learn the lessons from previous such experiences . . . [W]e have not institutionalized that knowledge; we have not integrated it into our doctrine, our training, and our planning for future operations. Neither have we regarded people with experience in this field as a national asset, to be retained, rewarded for good service, trained further and placed in positions from which they can be made available the next time such skills are called for."
[211] See: United States Department of Defense. Secretary of Defense Annual Report to the President and the Congress, 2003. 1 Dec. 2007. http://www.defenselink.mil/execsec/adr2003/adr2003_toc.html. This does not mean that U.S. troops are not capable of conducting non-combat tasks. Special Operations Forces (SOF) with specialised tactics, equipment and training, foreign language skills, and flexible unit deployment options, for example, are tailored to a wide range of tasks. Given their linguistic, cultural, and political training, SOF are well suited for both coordinating humanitarian assistance operations and performing combat search and rescue missions.
[212] Missiroli, Antonio. "Mind the Gaps: Across the Atlantic and the Union." Shift or Rift: Assessing US–EU Relations after Iraq. Ed. Gustav Lindstrom. Paris: EU Institute for Security Studies, 2003. pp. 77–90.
[213] Gettleman, Jeffrey. "The Struggle for Iraq: The Troops. American Commanders Say that Keeping a Lower Profile in Baghdad Is Their Primary Goal." New York Times, 10 Feb. 2004.

7.3 The global and European security context: a proposal for space

Erwin Duhamel

7.3.1. The global and European context

This article provides a detailed description of the actual and potential contribution of space assets to security policy. Following a look at the global and European context, it assesses the perspective of Europe's engagement in this field. Then it details the use of space systems and finally focuses on the aim and scope of a space doctrine. The viewpoint for the following analysis is European, but the context is certainly global.

The worldwide security environment has undergone fundamental changes in recent years. The main threats of today have a completely different shape, direction, and pace than those of the past. At the same time, the European Union has evolved into a global actor, increasingly called to playing a major role in global politics, including the field of security.

7.3.1.1. The global security context

The *United States* is today's largest space power and the revised U.S. National Space Policy issued in 2006 focuses explicitly on the "freedom of action in space" in support of homeland security. In 2006, *Russia* adopted a new 10-year Federal Space Programme and plans to progressively raise the funding levels in support of the modernisation of its existing space systems, many of them serving both civilian and military purposes (e.g. GLONASS). *China* has delivered proof of successfully mastering all space technologies, including staffed transportation systems and ASAT capabilities, and is developing an important number of scientific and application satellites for Earth observation, telecommunications, and navigation apart from launchers. An increasing number of nations, including *Japan*, *India*, *Brazil* and others, are committing significant efforts to obtaining space assets that also serve security purposes.

7.3.1.2. The European security context

The evolution of Europe's role in security and defence has been accelerated by political decisions taken immediately after the 1998–1999 Kosovo war. These decisions subsequently paved the way for a series of policies relevant for the European Union's security and defence activities in general and, later on, for the role of space in particular.

7.3.1.3. The European Security and Defence Policy

The European Security and Defence Policy (ESDP) is the operational security component of the Common Foreign and Security Policy (CFSP) of the European Union.

7.3.1.4. The European Space Policy

The ESP stipulates, as a key action in the area of security and defence, the establishment of a mechanism to exchange information and identify opportunities for increasing coordination and synergy between the different actors.

7.3.2. A possible future for the European use of space for security

Until now, the European military space programmes have been developed more as "individual projects" than as elements of an (even partly) integrated infrastructure based on a global vision. To foster this "global vision" and thus ensure an optimised level of coherence and connectivity between future capabilities, some fundamental documents are required:

- A European doctrine on the use of space for crisis prevention and management, including combat operations;
- The establishment of concepts of operation for each type of space capability;
- The specifications of a global architecture to be implemented incrementally.

Based on the ESP and the ESDP, this article took as a basic hypothesis a common understanding of operational needs. These needs must still be prioritised and converted into detailed operational specifications. This article does not reflect the point of view of ESA. It is only a DG-

PS reflection that could serve as a basis for data policy and governance discussions.

The key principles to be applied to the design of a security space infrastructure include interoperability, the pooling and sharing of assets, and responsiveness. In order to improve performance at all levels, an appropriate compatibility and connectivity policy must be developed within the framework of a European Network Centric Action, to interconnect all assets, in particular communications equipment including terrestrial, airborne, maritime and space-based equipment. Important goals to be achieved are increased situational awareness and a common operational picture.

7.3.2.1. Information-gathering and intelligence elaboration

Information-gathering and intelligence elaboration are a pre-requisite for the EU to achieve autonomous situation awareness and decision-making capability. Furthermore, information and intelligence sources are absolutely mandatory for any crisis prevention and management activities. In particular:

- Before the occurrence of a crisis, continuous intelligence derived from information gathered notably from space assets supports the political assessment of the situation and helps political leaders to deter the hostile party from entering a real crisis; furthermore, it creates the availability of documented information ready for use.
- During the crisis management phase, information gathering from space in a timely manner is a highly valuable tool for:

  - The planning and
  - Assessment of adverse activities,
  - The preparation, deployment and coordination of military forces and civilian aid workers,
  - Search and rescue operations,
  - The acquisition of infrastructure status,
  - The evaluation of the success of operations,
  - Tracking the movements and the displacement of refugees.

- In the framework of post-crisis operations, information-gathering allows the monitoring of:

  - The implementation and/or the respect of peace/cease-fire agreements,

- o Demobilisation and disarmament operations,
- o The tracking of the resettlement of displaced populations.

- Furthermore, information-gathering is of great interest for ensuring compliance with international treaties (proliferation, disarmament).

Space contributes to information-gathering, which encompasses different aspects such as:

- The overall political, diplomatic, ethnic, religious, cultural and economic situation,
- The actual geographical environment including border surveillance, types of deployments, locations and activities of sensitive infrastructures, possible support and/or constraints linked to the populations, local resources,
- The capacities of hostile forces, including doctrine, combat structures, procedures and the performance of their security assets,
- The actual forces' volume, combat structures and capabilities, police forces, militia, terrorists, populations/refugees,
- The situation linked to humanitarian and rescue tasks,
- The overall or specific situation to support the EU and third countries in combating terrorism.

7.3.2.2. Command and control, including CIS

Command and control (C2) must be ensured at any given time as soon as the action/involvement of the EU has been decided. This requires seamless and interoperable CIS support, able to transfer and share globally a large amount of highly classified information. The whole chain of command and control, from political strategy to the tactical level, must rely on networks supported by ready-to-use (plug-and-play) and robust, rapidly deployable assets.

7.3.2.3. Positioning–Navigation-Timing (PNT)

Most modern weapons systems rely on highly accurate and reliable positioning space capabilities in order to deliver their focused and effect-oriented power with minimum collateral and environmental damage or casualties. These space capabilities must be operated in a highly secured environment. They must remain under the control of EU authorities. Furthermore, the efficiency of relief aid and forces is strongly enhanced by the real-time knowledge of their position, especially in areas where maps do not exist or are inaccurate or obsolete.

The increasing use of ciphered telecommunications and digital systems requires an efficient and accurate network synchronisation and the distribution of a common precise time reference. Within this framework, PNT space systems are essential.

7.3.2.4. Protection of national assets

Any potential hostile entity may gain the capability of gathering information relating to any sensitive infrastructure in EU member states in areas such as industry, transportation, or energy distribution. The future EU space infrastructure for ESDP purposes must take such growing risks into account.

In addition to the usual protection procedures by national or collective means, a global protection of national infrastructures is required against the potential use of hostile intelligence gathering. The evolution of threats associated with terrorism implies the protection of the national territories of the member states. Space assets must actively contribute to the global protection concept, which encompasses:

- A threat detection and characterisation function, aimed at providing the EU and the populations with sufficient notice to take appropriate national or collective protection actions;
- Means of passive defence: counter-measures ranging from camouflage, concealment, deception, hardening, anti-jamming, anti-spoofing and anti-intrusion to information security in a broad sense;
- Means of active defence: blinding, active jamming, disabling of enemy ground segment capabilities;
- Damage assessment, consequence management, graceful degradation.

7.3.2.5. Responsiveness

The ambition of the EU to act quickly implies the need to:

- Permanently rely on basic pertinent and up-to-date information;
- Have the capacity to increase intelligence capabilities in the area(s) concerned on very short notice; this can be achieved either through a different and more focused use of already existing assets or through the addition of new ones;

- Have access to ready-to-use communications means as a rapid support to the architecture of the required networks;
- Deploy aid through the available transport means.

7.3.2.6. General support

In addition to the aforementioned needs, the conduct of operations requires capabilities such as:

- Timely-updated maps, including general and specific cartography;
- Imagery to rapidly provide up-to-date information, to prepare specific tasks, to obtain data that allow factual analyses to be performed, to answer legal enquiries, and more generally, to obtain evidence that reinforces the legitimacy of EU actions;
- The synchronisation of systems on the ground;
- Meteorological and sea state forecasts, including climate knowledge;
- Real-time environmental information and forecasts (environmental hazards).

7.3.3. Space systems for security

Satellites orbit in a free environment where sovereignty constraints cannot apply in accordance with international law, as stipulated by the 1967 Space Treaty. They are less vulnerable than other assets and provide their operators with a global access capability following controlled and predictable patterns, with a high degree of autonomy, relative discretion, and repetitively. They can be tasked at short notice when necessary. They create a very low direct risk for the populations and the environment.

Since the beginning of the 1990s, the effective use of space capabilities has contributed to information superiority and effectiveness. Space offers an excellent position from which to communicate, observe and listen. Space assets are therefore a multiplier to the benefit of the ESDP. In the telecommunications area, it allows the immediate dispatching of important information flows over large (distant) areas, anywhere in the world, in quasi-real time, and requires less ground-user infrastructure than traditional terrestrial networks.

The specificity of space techniques in general requires the permanent availability of relevant specialists (technical experts and operators) in the member states who have to be regularly trained at both the national and European level to be prepared for acting more efficiently during crisis management.

"Space systems" include both space-based capabilities and their respective ground segments. The overall efficiency of a space infrastructure strongly relies on the interoperability of individual space (sub-)systems, including their ground segments and their interfaces with the users.

7.3.3.1. Contributions of space systems to the ESDP

Space systems are by no means the only capability able to support ESDP operations. Due to their specific advantages, space systems are complementary to traditional assets and platforms. Aircraft and UAVs cannot legally and safely fly over potential enemy territories, whereas a reconnaissance satellite can perform this without any risk. This preliminary remark is also valid for other space applications such as telecommunications, SIGINT or early warning systems. More generally, as "information providers and dispatchers", space systems are undisputed components of network-centric infrastructures. A series of requirements are common to all space capabilities:

7.3.3.2. Space system responsiveness

Responsiveness must become the key parameter in the design of all future space capabilities. The current security space systems remain built and operated as if they still had to meet strategic requirements attached to a 'classical' war threat. To cope with the new threats (terrorism, unpredictable crises . . .), future systems will need to be designed and optimised not only at the strategic level, but also at operative levels, which remains a real challenge. The anticipated responsiveness of space systems ranges from a few minutes (early warning) to a few hours (meteorology, operative imagery, ELINT, oceanography). The short response time required by European authorities to enable fast situation assessment and decision-making goes beyond technical solutions. It encompasses various issues, such as:

- The design of capabilities on the basis of practical requirements expressed by experienced staff in all areas of space;
- The creation of the core of the future European security space infrastructure in the form of a general purpose ground network which enables the secure interconnection of space assets and user facilities by means of standard formats and functional interfaces (plug-and-play);

- The timely exploitation and analysis of data;
- The optimisation of the space segments to truly meet the user ground segment requirements;
- The introduction of data relay communications services for timely sensor programming, data downloading and dissemination;
- The agreement of an information security policy encouraging the circulation of information at all levels;
- The development of user-friendly application software to enable the access to space assets by non-specialist staff;
- etc.

7.3.3.3. Space system security requirements

The future European security space infrastructure will have to protect all space assets by taking into account the full spectrum of well-known and innovative actions that may threaten both space and ground segments. Space systems are subject to various threats that can be classified as follows:

- Deception: the manipulation, distortion or falsification of information;
- Disruption: the temporary impairment of a capability;
- Denial: the temporary elimination of a capability;
- Degradation: the permanent impairment of a capability;
- Destruction: the permanent elimination of a capability.

The ground segments (industry facilities, launch bases, satellite and mission control centres, terrestrial communications networks, user stations . . .) need to be protected against traditional terrorist attacks (explosives . . .), cyber-terrorism (satellite and mission control software, data communications networks), electronic intelligence (telecommand and telemetry interception), data reception jamming, etc.

The space segments could be subject to attacks by ground-based and (eventually) airborne laser beams, microwave jamming, kinetic energy space weapons, proximity explosions, telecommand spoofing, cyber-attacks of on-board software, etc.

Adversaries, including terrorists, are able to perform a wide range of attacks on space systems (space and ground segments) using commercially available or more sophisticated equipment and weapons. The unprotected commercial communications systems used heavily during crisis periods could be affected. A comprehensive space system security policy covering the space and ground segments as well as space systems operations in general is thus needed. This policy should encompass:

- The protection of space segments: component hardening, on-board sensors to detect the attacks, aggressor identification means, increased autonomous operations . . .
- The protection of ground segments: against physical sabotages, anti-cyber attacks, anti-jamming, mobile TT&C stations . . .
- An information security policy covering electronic attacks on communications, data and command links – from operational data (telecommand and telemetry) to user level information acquired from space (imagery encryption . . .).

The security of space systems could be improved by the deployment of space situational awareness, ELINT and COMINT systems.

7.3.3.4. Space system interoperability requirements

When applied to space systems, the term "interoperability" may have different meanings. For telecommunications systems, interoperability means mainly "cross-compatibility". Future European satellite communications systems must meet the relevant international standards. Considering a given spectrum band, the standardisation of the access to European satellite communications systems could lead to significant improvements in capability, as well as to economies of scale. Another step in "interoperability" can be reached when security user terminals are able to provide access to all commercial systems.

For Earth observation systems (high-resolution 2D and 3D imagery, environmental data . . .), "interoperability" is needed at the level of the ground segments. The standardisation of processed data formats and access will be required to enable the easy fusion of data acquired from different space and non-space sensors. The access to various image data bases with the same terminal through a unique multipurpose network, and by using the same software interfaces – in particular the man–machine dialogue software – remains a priority for the end user.

For satellite navigation, "interoperability" will refer to supplying the forces with "dual-mode receivers" combining the processing of both GPS and GALILEO signals. The pooling and sharing of the above-mentioned and other capabilities (early warning, SIGINT, meteorology, oceanography, space surveillance . . .) will require an additional understanding of "interoperability" in the form of standardised access to a general purpose network by means of data presentation formats and INTERNET-like protocols to be mutually agreed upon by the European partners.

7.3.3.5. Space system autonomy requirements

The ESDP autonomy in space for crisis prevention and management purposes requires that the following key requirements be met:

- Guaranteed access to space, including launchers and fuel. It should be noted that the ARIANE 5 launch conditions for small satellites – a tendency in Europe – are not fully satisfactory from a security viewpoint. Even the "small" launcher VEGA currently under development will still be too "heavy" (and too expensive) for satellites of the 150–200 kg class or too "light" for polar orbiting satellites of 2.5–3 metric ton class;
- European autonomy in the field of space components remains to be established and maintained. Most sensitive or hardened components have to be imported and are subject to severe export regulations causing delays in the development of space systems;
- Finally, all space capabilities mentioned in this document should be seen as elements of a global architecture which contributes to the establishment of a European autonomy in space for security purposes.

7.3.3.6. R&D requirements

Most – if not all – existing military space capabilities and those under development in Europe have been derived from earlier pathfinder civilian programmes. But a number of the technologies and techniques required for addressing new threats do require a specific R&D programme for security and defence purposes. Among the possible space segment research and development topics for further study are the following:

- High data rate, wideband, multimedia communications, as well as multi-access relay communications techniques using microwave and optical technologies;
- Compact, extremely high resolution optical and radar imagery instruments, as well as automated data processing, fusion and analysis techniques;
- The characterisation of visible/infrared plurispectral and hyperspectral bands for environmental cartography specific to security and defence;
- Specific aspects of security systems (SIGINT, early warning, self-protection systems . . .);
- The miniaturisation and hardening of on board components (mechanical, electronic, electromechanical . . .);
- Spacecraft flight information techniques (precise relative position determination, accurate positioning control . . .);

The activities carried out under such a research and development programme should range from equipment pre-developments at ground to in-orbit demonstrations and qualification to enable the decision of operational programmes with the lowest possible risk.

7.3.3.7. Considerations on dual use issues

All future development of space systems should take into account their dual use dimension. Though this is obvious in areas such as telecommunications, imagery, navigation, meteorology or oceanography, the following areas are worthy of note:

- Early warning: required for missile launch detection and also needed for the monitoring and implementation of disarmament treaties by the relevant civilian authorities;
- Space situational awareness: required for the security actors and also needed for the civilian monitoring of space debris, solar activity (space weather), NEO (Near Earth Object) or atmospheric re-entries.

Any organisation established to prepare dual-use mission specifications must ensure that:

- The security requirements are fully integrated;
- The information security principles required can be fully implemented.

The interaction between the security and the civilian sectors appears at two levels:

- The specifications of the content of future research and development programmes, in order to avoid any duplication;
- The specifications and the implementation of dual-use, operational programmes such as Pléiades or other future systems.

7.3.3.8. Aim and scope of a Space Doctrine

Based on the "European Space Policy" and "ESDP and Space", the aim is to provide guidelines for the employment of space-related assets and capabilities by the EU during the early decision-making process, the planning and the execution of military and civilian crisis management tasks.

7.3.3.9. General considerations

The potential space capabilities to be deployed by the member states in support of ESDP operations will belong to one of the following categories:

(1) "Military space capabilities", i.e., space systems for which the design incorporates military constraints and access to which will be subject to military regulations;
(2) "Dual use space capabilities", i.e., civilian systems which incorporate certain military constraints. Access to such systems might thus be subject to military regulations;
(3) "Multiple use space capabilities", the data acquired of which can be used by both civil and defence communities. However, under crisis circumstances, access to such capabilities may be subject to temporary restrictions in order to prevent any misuse of the information provided.

7.3.3.10. Legal constraints

Establishing the ESDP Space Doctrine needs to take into account the legal framework and potential limitations. There are currently five UN Treaties and five major UN Resolutions that govern the activities in outer space. The internationally recognised legal principles originating from these UN texts can be summarised for the purpose of this document as follows:

- Non- appropriation of outer space by any country,
- No deployment of nuclear weapons or Weapons of Mass Destruction in outer space or on celestial bodies,
- The freedom of exploration and use of outer space for the benefit and in the interest of all countries,
- The liability of states for damages caused by their space objects,
- The safety and rescue of spacecraft and astronauts,
- The moon and other celestial bodies shall be used exclusively for peaceful purposes,
- The prevention of harmful interference with space activities and the environment,
- The notification and registration of space activities,
- The scientific investigation and the exploitation of natural resources in outer space and the settlement of disputes.

According to the currently prevailing interpretation, the term "peaceful" in clauses referring to "peaceful purposes" or the "peaceful use" of outer space means the "non-aggressive" purpose or use as part of space activities. In this context, it should be noted that the EU does not intend to implement any concept such as "space dominance" or the "weaponisation of space".

Recent space achievements have demonstrated that non-state or private actors now have access to advanced space technologies that allow them to accomplish missions that used to be restricted to governmental entities. In the longer term, such activities can only develop further and will finally generate new threats that need to be anticipated. Therefore, the member states are required to elaborate common export principles and regulations on space matters.

7.3.3.11. Technological constraints

The space research and development activities in support of the ESDP must rely on a formal selection process that takes into account a series of factors such as common operational requirements derived from long term mutually agreed threat assessments by the member states, specific operational requirements expressed by the national defence communities, the priorities with respect to available resources, and the space asset deployment plan. The high cost of investments and the relatively low quantities produced do not allow Europe to reach a complete level of independence. A worldwide co-operation effort involving partners confronted with the same threat should therefore be sought for.

7.3.3.12. Regulatory constraints

Europe needs alternative schemes for securing the procurement of sensitive space components and equipment that may be subject to tighter export regulations on which both military and commercial space activities rely.

7.3.3.13. Industrial constraints

Both the defence policies and the budget resources available in member states limit the number of space assets deployed. This results in a highly irregular workload within the space industry that is not offset by the commercial space market. Workforce, professional skills, technical know-how and ground industrial infrastructures require maintenance by appropriate mechanisms.

R&D programmes and mission demonstrators may improve the situation of the industry in the short term, but are not an ideal solution for ensuring the European industrial base. A permanent assessment of the industrial situation is thus required.

7.3.3.14. Space Doctrine implementation principles

Based on the foregoing, the implementation principles applicable to the full range of space assets and capabilities in support of the ESDP include the following aspects:

- Increasing the level of space awareness of the EU and its member states;
- *Overarching architecture:* the EU needs to establish a global architecture for the various components needed to build the defence space infrastructure, leaving open all options for its implementation;
- *Readiness and responsiveness:* a rapid reaction capability should be regarded as an inherent characteristic of the EU during CMOs. Therefore, access to space assets through pre-defined arrangements between the EU and capability operators should be secured;
- *Compatibility*: national space assets should apply the applicable EU standards in the future, and should be compatible with already existing networks and processing systems. An agreed common standard for future civil and defence-oriented space programmes will allow a dramatic limitation of the number of ground assets needed to receive and process data from all space-based sources.
- *Interoperability:* national space assets should, in the future, be able to work together within the EU network envisaged and with already existing networks and processing systems;
- *Connectivity*: inter-linking capabilities between various space assets such as data relay or between space and non-space assets must be aimed at;
- *Pooling and sharing:* the member states are invited to pool and share information collected from space-based and ground-based sources in order to allow more effective operational achievements within the framework of a European network-centric architecture. In order to ensure the best co-operability, the communication networks enabling access to such information will offer multi-facet interoperability, combining multi-level access control, multi-level secured communications, standard equipment interfaces, and harmonised access procedures. The interfacing of traditional information collection platforms with space-related assets is encouraged;

- *Technological independence*: in the development of critical elements within the space infrastructure in support of the ESDP, the extensive use of non-EU components and equipment should be avoided as far as possible;
- *Co-operation:* the principle of shared competence applies to the deployment of space assets in the EU. The provision of such assets is governed by various forms of binding acts between the partners. Such binding acts may be activated when a new space programme is decided, which allows an easier sharing of industrial work packages (even at a later stage of its development or deployment). The arrangements may foresee capability exchanges, contributions in kind or cash funding in exchange for appropriate capability access. They may also include specific arrangements to meet the EU requirements for crisis management operations, sharing their deployment costs. The conditions of access to such assets are subject to specific arrangements that may vary depending on the category which a particular asset belongs to: "Military Space Assets", "Dual Space Assets", or "Other Space Assets".

7.3.4. Implementing guidelines: space command and control

The EU space command and control (C2) in support of the ESDP should be executed by a Space Task Force (STF). Concerted action is essential for an efficient recourse to space assets in support of ESDP operations. The level of investments required for the deployment of space systems, the need for a European global vision of space assets for ESDP purposes, the intrinsic complexity of space systems, and the need for developing a common understanding of space systems use within the member states requires the creation of a dedicated instrument. To achieve the strength of a concerted space use, to optimise the use of scarce space assets, and to ensure that space capabilities are used as the overall situation commands, the following key principles must be recognised:

(1) The coherence and effectiveness of space use is most likely to be achieved when the command of space-related operations is exercised from the highest practicable level where the priorities of demands on those limited assets can be best assessed. Only a Space Task Force under the responsibility of the High Representative of the Union for Foreign Affairs and External Relations or of the Permanent President of the European Council is in a position to balance the changing and often conflicting requirements of space capability use. Thus, the Space Task Force should be the single focus for EU Space C2, within its terms of assignment.

(2) The Space C2 set up in support of the ESDP should apply to the full range of space capabilities. The potential “multiple use” of products and services of “space assets” is generally endorsed by the civilian and military communities. However, access to civilian and commercial space assets may be subject to ad hoc restrictions relating to national sovereignty issues, crisis management needs, and vital interests of the EU and its member states.
(3) Concerted efforts between European partners will be required for ensuring the maximum benefit of space assets. This includes key issues such as the overall system architecture and coherence, space mission objectives and schedules, R&D, evolving technologies, space component policies, space system interoperability and responsiveness, standard capability access interfaces and procedures using low-cost user equipment, and capability-funding mechanisms. Historically expressed in terms of missions, capabilities and sensor performances, the operational needs should progressively be converted into “services”;
(4) The use of space assets is an indispensable element for the protection of national forces and assets. In some cases, only space-based information collection and analysis can lead to the implementation of appropriate national and collective defence measures. The successful implementation of such protective measures implies that both the space and ground segments of the European space architecture are easily accessible, reliable, responsive, and close to the end-users;
(5) In order to have the maximum benefit from of space use, the Space Task Force will have to interact with highly trained personnel of the member states that are capable of operating in a CMO environment and offering the full range of skills: technical, operational, analytical, and decision-making.
(6) The execution of Space C2 tasks under the supervision of the Space Task Force will mainly rely on the support of external entities such as the European Defence Agency, national military entities, the European Space Agency, the EU Satellite Centre, national space agencies, consultancy firms, commercial service providers, etc.

The Space Task Force will be responsible for the preparation of the short and long-term actions required in support of the ESDP. The staffing will consist of representatives from member state forces mastering the space use for CMO. As a minimum, the following tasks need to be included:

(1) Establishment of a Global Space Vision. In order to prevent a persistent shortfall situation concerning the availability of space capabilities for the ESDP, the Space Task Force is in charge of maintaining a global vision of space (ESP) and ground product and service requirements for the ESDP. This

vision includes the full spectrum of space and ground related capabilities constituting the space infrastructure to support the ESDP, long-term implementation schedule needs, as well as various options to be considered for the implementation of such an architecture. This vision should be elaborated in consultation with the entities willing to provide access to their space assets and other bodies as appropriate;

(2) Elaboration of Space Plans. The STF will carry out analyses of the use of space assets required at all levels. The space capabilities provided by member states or organisations must be matched against these requirements. The main selection criteria will be the interoperability provided by these systems to support CMO C2. The identification and prioritisation of relevant space resources and the request for the necessary core space support must remain a key objective. Space plans should include the use of these assets for bridging shortfalls in accordance with the requirements;

(3) Ensuring efficient space and ground capability interaction. This requires to convey and co-ordinate data acquisition and to manage the information exchange process between the space and ground elements as well as the distribution among the EU and all member states in order to avoid overlapping or shortfalls. The STF should also promote the future EU architectural structure as a basis for the interoperability of future EU members' space system and ground control segments;

(4) Maintaining operational situation awareness on space-related resources, including:
   - Operational space requirements, i.e. space-related products and services potentially required by the EU for crisis prevention and management operations. In particular, the STF will collect present and future needs from the strategic level to the tactical one,
   - The full range of space assets and services potentially accessible by the EU, as well as their respective conditions of access. The STF will ensure the global coherence of space use for the ESDP,
   - Quality assurance, performance monitoring and, more generally, return-on-experience. The STF will thus directly contribute to space use improvements for ESDP purposes;

(5) Elaborating the required detailed guidelines for framework arrangements (including co-operation principles) with potential product and service providers, which can be:
   - Any entity designated by the EU to operate EU-funded space systems,
   - Any entity designated by member state governments to operate government-funded space systems,
   - Any entity operating a space system on a commercial basis;

(6) Initiating a concerted education and training programme on space use for ESDP in co-operation with the EU and its member states. Such a programme will progressively foster the dissemination of space skills among the EU and its member states, facilitate the exchange of experience between national space users and, hopefully, increase EU and member state commitments for the development of future space-related capabilities.

7.3.5. The way forward/What lies ahead?

While hesitant to the establishment of supranational European security and defence capabilities, space infrastructures put in place by national defence or dual-use programmes are *de facto* foundations on which the space segment serving a European Security and Defence Policy can be built.

European collaboration takes place in a multidimensional way. Figure 1 shows the collaboration scheme used in the Helios II programme: France and Germany exchange Helios and SarLupe images, Italy and France exchange Helios and COSMO-SkyMed images.

In December 2006, France, Germany, Belgium, Spain, Greece and Italy signed an arrangement related to initial collaborative studies for the definition of a future

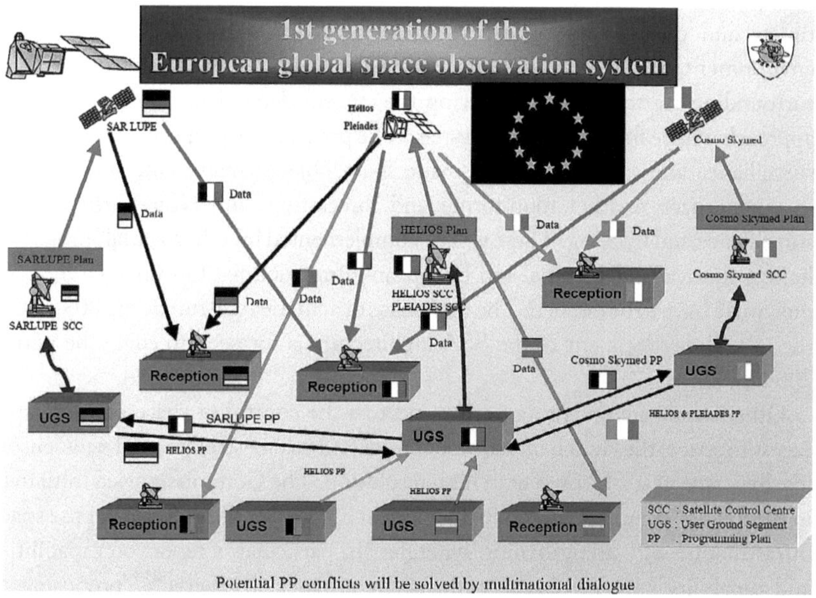

Fig. 1. *First generation of the European global space observation system (source: CNES).*

space-based system for surveillance, reconnaissance and observation missions called MUSIS (Multinational Space-Based Imaging System).

The signature of this arrangement is the first step towards the establishment of a space-based surveillance, reconnaissance and observation system which will succeed the HELIOS II and SAR-Lupe military systems and the COSMO-SkyMed and Pléiades dual systems.

The work to be performed will deal on the one hand with the definition of a global architecture to answer the capacity need in the field of Earth imagery for surveillance and reconnaissance purposes. The architecture that will be selected at the close of the study will become the basis on which the future optical and radar satellite programmes will be launched. On the other hand, the work will deal with the definition of a multi-sensor user ground segment which will allow each country to have access to all the satellites of the future system through a unique entry point.

This arrangement is an extension of the agreements concluded between the nations on the HELIOS II, SAR-Lupe and COSMO-SkyMed systems. It gives evidence of the will of the six countries to continue and reinforce their collaboration in the space field by together paving the way for a future space-based independent European surveillance, reconnaissance and observation system. This initiative marks an important step forward in the building of a European defence system.

The Space Situational Awareness (SSA) initiative will support the independent European utilisation of, and access to, space for research or services by providing timely and quality data, information, services and knowledge regarding the environment, the threats and the sustainable exploitation of the outer space surrounding Earth, thereby removing the current dependency on the U.S. The approach will be for a distributed system. The programme proposal comprises: the surveillance and tracking of man-made space objects; man-made space objects imaging; space weather monitoring and forecasting; and Near-Earth Objects surveillance and tracking. These will be complemented by a "horizontal" project on linking the available national and European infrastructures. Governance and data policy will have to be defined. The first phase of activities will run from 2009–2011; the overall development of the SSA architecture is foreseen to cover the period 2009–2018.

Other needs are emerging with respect to the control of space such as rapid access to space, the launch on request of mini and micro-satellites and new sensors like hyperspectral TIR ones with high resolution. The Commission-led initiatives in the field of security research further aim at identifying further gaps in the space infrastructure and strengthening synergies. In particular, a report on capabilities and capability gaps has been produced by a panel of experts,[214] proposing to conduct a series of assessments (e.g. a capabilities catalogue), technology development and interoperability projects to enable a series of services that space assets can provide in support of a more secure Europe.

European cooperation will need to be organised for the future generations of telecommunications and observation satellites. However, the process of reflection on such cooperation should start as soon as possible and focus on achieving systems interoperability. The challenge will be to take actions forward under a coherent approach involving all stakeholders. In this context, the establishment of a European Security Research Programme and the concerted action of the European Defence Agency, the European Commission, the European Space Agency and national agencies under the umbrella of the European Space Policy will be crucial in achieving European strategic objectives.

In order to resolve the questions of datapolicy and governance that arise in every European space project, the setting-up of the "Space Task Force" would solve a lot of problems. A forum where all actors involved are represented should be created to follow up on this idea.

[214] European Commission. Report of the Panel of Experts on Space and Security. Brussels: European Commission, March 2005. http://ec.europa.eu/comm/space/news/article_2262.pdf

7.4 Space for security and security in space

Gérard Brachet

7.4.1. Introduction

Human activities in outer space and the exploration of the solar system are now 50 years old. During these 50 years, space activities have developed tremendously in scope, and the number of actors involved has rapidly increased. Today, 10 countries master satellite-launching technologies and the operation of launch systems, and more than 50 states have satellite programmes. The commercial sector has also advanced significantly, with several private companies providing launch services and many more companies operating commercial spacecraft and associated services.

Looking at the founding motivations for space activities, one can easily identify three principal objectives:

- Scientific research and exploration,
- Applications for society's needs,
- Satisfying requirements in the security and defence area.

Each of these objectives plays a role in a country's decision to invest in space activities: While the initial impetus for developing space technologies came mostly from strategic considerations in the USA and in the USSR in the 1950s, scientific research soon became another major impetus. For example, in Europe, scientific research was a prime motivation for the establishment of the European Space Research Organisation (ESRO) in the 1960s, later to become the European Space Agency. Direct applications of space systems for society's needs emerged in the 1970s with telecommunications, meteorological and land observation satellite systems. Today, while science and exploration are still a major motivation for space activities, applications for the daily needs of human societies and for sustainable development purposes have become the main incentives for emerging states to invest in space systems. Security is obviously one area of application of direct interest to these states. In addition, gaining access to the advanced technologies of the spacecraft sector and its associated ground segment as well as to the skills needed to operate complex systems with a global outreach is a prime motivation for many countries.

Thus, space systems contribute to overall security, but "space security", i.e. the ability to use space in a sustainable manner, is also a question that we need to address because of the tremendous increase in the number of orbiting spacecraft, of the scarcity of frequency resources in the electromagnetic spectrum, and of the fast growing amounts of debris in the near-Earth orbit.

7.4.2. Security in space

Looking back, looking forward, the international legal framework for space activities and the future outlook for a secure space environment 10 October 2007 marked the 40th anniversary of the entry into force of the Outer Space Treaty (OST), signed by the principals in London in January 1007 and ratified in the same year. The OST and the successive treaties and conventions related to space that were negotiated and entered into force in the 1970s and 1980s provide the international legal framework for space activities worldwide (Table 1).

In particular, the OST includes key features such as the freedom of exploration and utilisation in outer space (Article 1) as well as the non-extension of state sovereignty into outer space and the non-appropriation of space or celestial bodies (Article 2). It also prohibits the deployment of weapons of mass destruction in outer space. Theses treaties and conventions were negotiated during the cold war

Tab. 1: *International treaties and conventions elaborated by UNCOPUOS.*

1967: Treaty on Principles Governing the Activities of States in the Exploration and Use of Outer Space, including the Moon and Other Celestial Bodies (entered into force the same year);
1968: Agreement on the Rescue of Astronauts, the Return of Astronauts and the Return of Objects Launched into Outer Space (entered into force the same year);
1972: Convention on International Liability for Damage Caused by Space Objects (entered into force the same year);
1975: Convention on Registration of Objects Launched into Outer Space (entered into force in 1976);
1979: Agreement Governing the Activities of States on the Moon and Other Celestial Bodies (entered into force in 1984 but signed and ratified by only 13 countries).

and some of their features reflect this historical background, but for the most part, they are considered adequate for today's space activities.

In addition to the body of international treaties and conventions, a number of principles and/or guidelines have been developed by the UN Committee on the Peaceful Uses of Outer Space (UNCOPUOS), established by the UN General Assembly in 1959, and were adopted as UNGA Resolutions (Table 2). One good example are the "Remote Sensing Principles" adopted in 1986 (UNGA Resolution 41/65) that spell out the principles to be followed for satellite-based remote sensing activities on the Earth's surface.

In 2007, UNCOPUOS adopted the "Space Debris Mitigation Guidelines", based on the excellent work done over many years by the international Inter-Agency Debris Coordination Committee (IADC) which consists of experts from thirteen major space agencies. These guidelines were adopted by the COPUOS Scientific and Technical Subcommittee in February 2007 and by the plenary Committee during its 50th session in June of the same year. They were later

Tab. 2: *UNCOPUOS Resolutions adopted by the UN general assembly.*

- Declaration on Legal Principles Governing the Activities of States in the Exploration and Use of Outer Space (1963), precursor to the Outer Space Treaty;
- Principles Governing the Use by States of Artificial Earth Satellites for International Direct Television Broadcasting (1982);
- Principles Relating to Remote Sensing of the Earth from Outer Space (1986);
- Principles Relevant to the Use of Nuclear Power Sources in Outer Space (1992);
- Declaration on International Cooperation in the Exploration and Use of Outer Space for the Benefit and in the Interest of All States, Taking into Particular Account the Needs of Developing Countries (1996);
- Resolution 59/115 of 10 December 2004 on the concept of "Launching State";
- Resolution 62/101 of December 2007 on the registration of space objects;
- Endorsement by General Assembly Resolution 62/217 of 21 December 2007 of the "Space Debris Mitigation Guidelines" approved by UNCOPUOS in 2007.

endorsed by the UN General Assembly in December 2007 (UNGA Resolution 62/217).

It is worth noting that among these Space Debris Mitigation Guidelines, guideline number 4 states: "Avoid intentional destruction and other harmful activities". It is hoped that the recent adoption of these guidelines by the UN General Assembly will discourage actions such as the Chinese Anti-Satellite (ASAT) "experiment" of 11 January 2007 which produced more than 2,200 additional pieces of debris larger than 10 cm, most of them long-lasting, in sun-synchronous orbit. The adoption of the "Space Debris Mitigation Guidelines" is an excellent example of the UNCOPUOS's approach to developing a consensus-based set of "Rules of Good Behaviour" aimed at minimising the production of space debris and the risk of collisions in outer space.

Another illustration of this pragmatic approach is the work plan for nuclear power sources in space. The UNCOPUOS Scientific and Technical Sub-Committee's Working Group on the Use of Nuclear Power Sources in Outer Space has established, in cooperation with the International Atomic Energy Agency (IAEA), a 3-year work plan which will lead to the adoption of a NPS safety framework by 2010.

Beyond the space debris issue, some countries are calling for a new treaty banning the deployment of weapons in outer space (only the deployment of weapons of mass destruction is forbidden by Article IV of the OST) and for a new comprehensive treaty replacing the five treaties already in force. However, there is no consensus on reopening the Outer Space Treaty of 1967, or to develop new international conventions. On the other hand, based on the UNCOPUOS work on space debris, there is a potential for a bottom-up, technically-based approach to developing guidelines and recommendations aimed at keeping outer space as safe as possible. Indeed, the increasing number of spacecraft in orbit, particularly in near-Earth orbits, and the increasing number of actors conducting space activities (both government and private operators) call for a fresh look at the long-term outlook for a safe space environment.

In June 2007, during the 50th plenary session of UNCOPUOS, the Committee addressed its future role and activities.[215] One suggestion I put forth in my capacity as Chairman of the Committee was to consider the development of "best practices" to ensure safe space operations for all actors in the long term. The recommended approach is to rely on the actual operational experience of the principal actors, commercial operators, and government agencies.

Informal consultations have been taking place with key countries since the summer of June 2007 to refine the terms of reference and a work plan for a future Working Group to be established within UNCOPUOS. This was the purpose of the informal working meeting convened by France that took place in Paris on 7 and

8 February 2008. Twenty states plus the European Union and the European Space Agency were represented, as well as four international governmental organisations. In addition, four major telecommunications satellite operators participated in the second day of the meeting. The participants agreed to prepare an outline document to be submitted to UNCOPUOS in 2009.

Thus, UNCOPUOS could gradually begin to play a key role in establishing a set of "best practices" or "Rules of Good Behaviour" for space activities. However, for these principles to be realistic and robust over time, they should be based primarily on technical and operational considerations and not on political considerations. It should be noted that this bottom-up approach remains firmly based on the Outer Space Treaty of 1967: it maintains the founding principles of the freedom of space exploration and utilisation (Article 1) and of the non-appropriation of outer space and celestial bodies (Article 2).

Turning to the military uses of outer space, UNCOPUOS does not explicitly address the military aspects of space activities, but its mandate covers both (non-aggressive) military and civilian uses of outer space. Besides, many space systems are "dual use" and it is often impossible to distinguish between their civilian and military aspects. Global navigation satellite systems such as the Global Positioning System (GPS) operated by the United States and GLONASS operated by the Russian Federation are typical examples of satellite systems which were initially deployed primarily for military uses and have gradually become widely used for civilian applications. Conversely, commercial telecommunications satellite services are used by the defence forces of many countries. In addition, military and civilian space systems share the same medium, the outer space environment, which cannot be divided into separate regions for military and civilian use.

The more specific issue of the potential deployment of weapons in outer space were discussed at the Conference on Disarmament (CD) in Geneva under the "Prevention of an Arms Race in Outer Space (PAROS)" agenda item. Recently, on 12 February 2008, the Russian Federation and China officially introduced a proposed treaty for the prevention of the deployment of weapons in outer space at the CD.[216] At the same time, the European Union has been working on a joint paper on Transparency and Confidence-Building Measures (TCBMs) which, once finalised, is to be presented to both the CD in Geneva and UNCOPUOS in Vienna.

Finally, it should be noted that in addition to UNCOPUOS and the Conference on Disarmament, the International Telecommunication Union (ITU) also plays a major role in facilitating the uses of outer space by organising and coordinating frequency allocations to all spacecraft, launchers, satellites, and staffed orbital stations. In addition, it allocates orbital slots in the geostationary orbit.

7.4.3. Space for security

Today, space plays an increasingly visible role for security on our planet Earth, at a local level as well as at a more global level. The word "security" must here be interpreted in its wider sense, i.e. as referring to the security of humans faced with adverse natural conditions (drought, effects of climate change, natural disasters), and to the prevention of, and protection against, human-made industrial and environmental situations affecting the safety and well-being of people. The concept of security also includes the prevention of, and protection against, acts of terrorism and other threats to the stability of human society as a whole.

Therefore, the many contributions of space technology and space-based systems to security can be considered under many headings, such as:

(a) Civil protection
(b) Environmental security
(c) Food security
(d) The prevention of conflicts
(e) Support for peacekeeping operations.

There are many concrete initiatives to use space systems for increasing security. I would like to present a few examples here:

In the area of natural disasters, the International Charter "Space and Major Disasters", initiated by the French Centre National d'Etudes Spatiales (CNES) and the European Space Agency (ESA) at the UNISPACE III conference held in Vienna in 1999, now includes 25 space agencies as partners who contribute their space assets to the collection of up-to-date Earth observation imagery whenever there is a major natural or industrial disasters and to the production of maps of the affected areas and other information of direct relevance for the rescue teams in the field. The "UNOSAT" team based in Geneva, which is part of the UNITAR organisation, plays a key role in producing these urgently needed documents and in forwarding them to the appropriate relief organisations irrespective of whether a national civil protection agency or a UN organisation, such as OCHA (Office for the Coordination of Humanitarian Affairs) or the HCR (High Commissioner for Refugees). Figure 2 presents a map of the northern part of Sumatra (Indonesia) following the devastating impact of the tsunami of 26 December 2004, and Figure 3 presents an image map of the city of New Orleans after it was struck by hurricane Katrina in late August 2005; these examples illustrate the type of document derived from satellite imagery collected soon after major disasters.

The International Charter "Space and Major Disasters" is now complemented by the newly initiated UN SPIDER programme, approved by the UN General

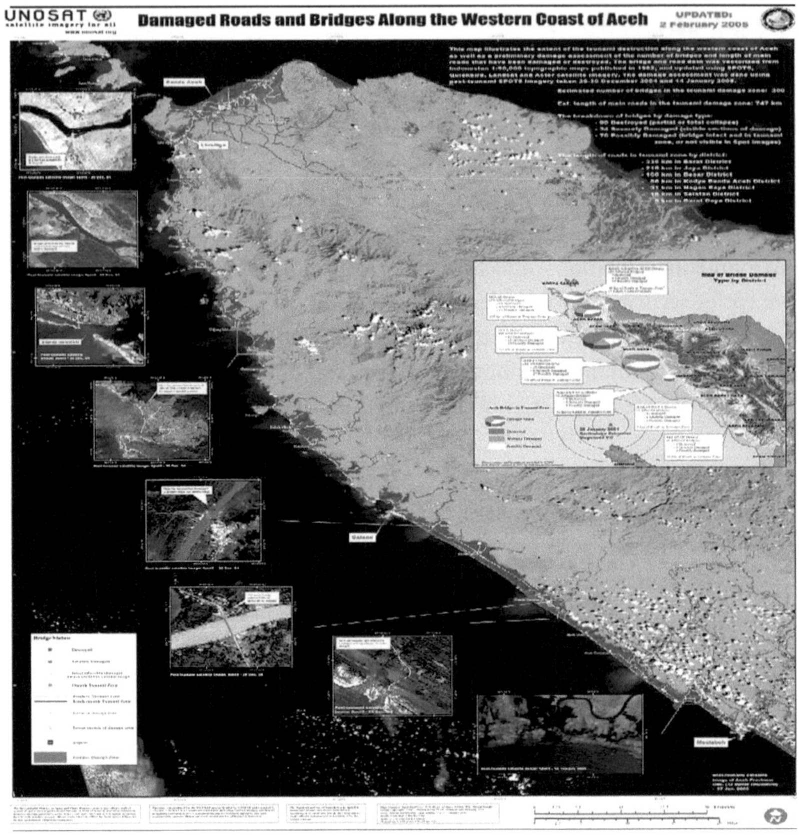

Fig. 2. *Map of the northern Sumatra region showing the coastal areas affected by the tsunami of 26 December 2004 (source: UNOSAT).*

Assembly in 2006 (Resolution 61/110) and managed by the UN Office for Outer Space Affairs which will facilitate the access of developing nations to space data and services and provide support in disaster situations.

The International Charter and the UN SPIDER programme tend to focus mostly on the collection of satellite-based imagery and their quick delivery to the concerned parties, but space systems can also provide very efficient and robust communication and geo-positioning resources. Clearly, one way to improve the use of space technology in the context of disaster or crisis situations will be to facilitate access to low-cost and robust communication terminals by the teams operating in the field. ITU has a programme focusing on such telecommunication applications and it is hoped that the UNOOSA-led SPIDER programme will be able to benefit from it.

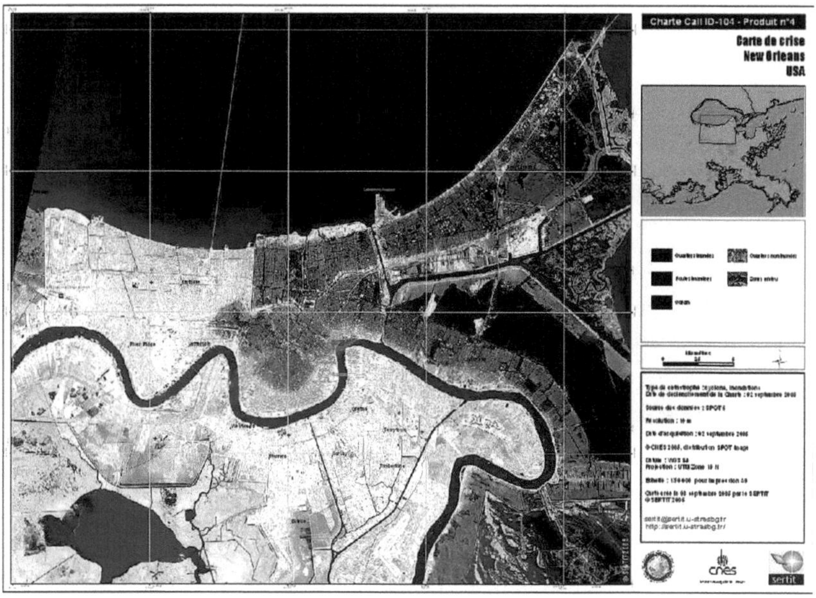

Fig. 3. *A crisis image map: New Orleans flooding based on imagery captured on 2 September 2005 (source: UNOSAT).*

In the field of space systems applied to the safeguarding of lives, space technology has made a major contribution to search and rescue operations in the last 25 years: the COSPAS-SARSAT system initiated in 1979 by Canada, France, the Soviet Union (now Russia), and the United States started operating in 1982. The system is composed of receiver/processors on board a series of low orbiting satellites and of relays on board of geostationary satellites. 38 nations are now participating in the system via dedicated ground stations and facilities. As of 1 January 2007, 22,412 lives have been saved thanks to this system since the end of 1982 (1,181 in the year 2006 alone).

Figure 4 illustrates a rescue operation at sea following a distress signal relayed by COSPAS SARSAT satellites, leading to a coastguard helicopter being dispatched to rescue the crew of the cargo ship.

In the field of food security, the contribution of space systems is twofold:

(1) Meteorological satellites form a major part of the meteorological observation networks that produce the input data for the very sophisticated numerical forecast models of the atmosphere that allow weather forecasts 5–7 days ahead for all regions of the world and, more recently, seasonal forecasts of climatic conditions for most regions. These forecasts are an indispensable input to the various national, regional, and global programmes aimed at ensuring that

Fig. 4. *Rescue at sea following a distress signal relayed by the COSPAS-SARSAT satellites (source: CNES).*

agricultural practices respond to the basic needs of the local population, while protecting the increasingly scarce soil and water resources.

(2) Land observation satellites produce imagery used for improved land management, for a better allocation of water resources to irrigation, for the early detection of stress factors affecting crops particularly in developing countries, and for crop production estimates. Figure 5 illustrates how satellite imagery of the same crop production area taken four times during the growing season enables the generation of estimates of crop acreage as well as the monitoring of crop growth.

In the area of international security, it is well known that satellite systems had a major stabilising effect during the Cold War, as they provided an unprecedented degree of transparency to the major players, namely the USA and the USSR. Imaging satellites provided them with detailed information of the military facilities being built or operated by the other party. In addition, early warning satellites in geostationary orbit capable of detecting the take-off of ballistic missiles were key elements of the nuclear deterrence strategy of the super-powers. This helped to maintain the equilibrium which produced the "cold peace" that characterised the world from the end of the Second World War to the early 1990s.

Today, space systems continue to play a key role in the security and defence strategy of major nations and are no longer limited to the two major players of the cold war. In the last 15–20 years, China, France, Germany, Israel, Italy, Japan, Spain and the United Kingdom have deployed military communication and/or reconnaissance satellites and are also considering deploying early-warning and eavesdropping satellite systems.

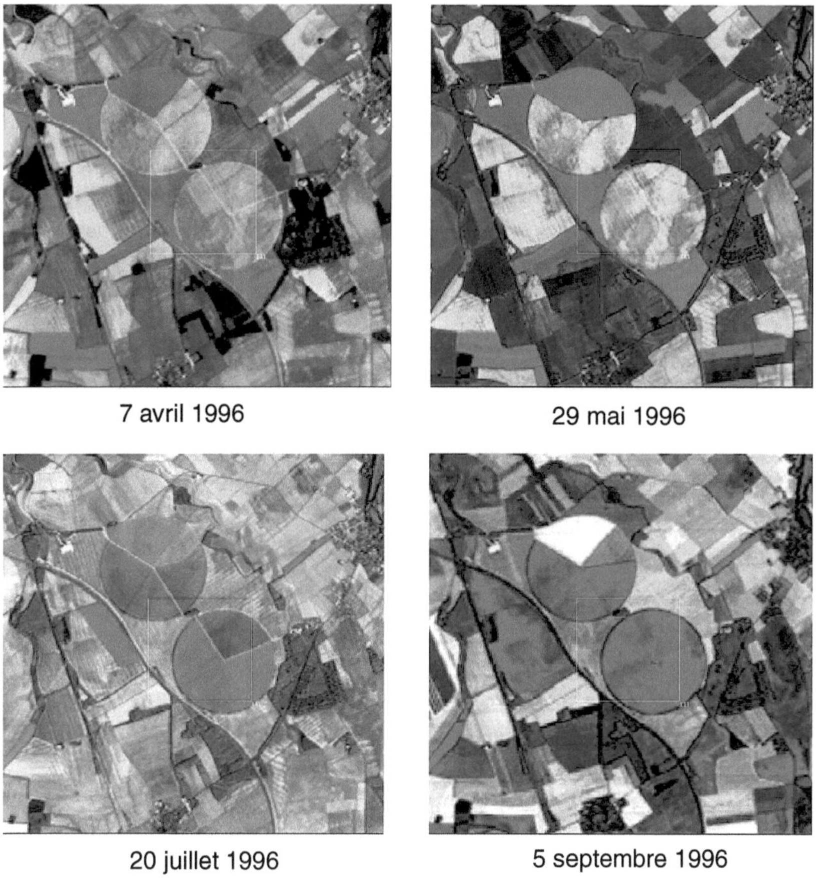

Fig. 5. *Monitoring of crops in a major agricultural production area in the region around Bourges (Central France), 1996 (source: Spot Image/CNES).*

7.4.4. Conclusions

Space exploration and the use of space in the 21st century will continue to be an exciting story, just as in the last 50 years. Space activities have very unique characteristics: a mixture of major strategic considerations, of scientific discoveries, of technological challenges, and contributions to our daily life and to sustainable development. These contributions to human society play a significant role in ensuring the security of ordinary people.

As space systems already contribute essential elements to the overall security of our societies on planet Earth, the future evolution of technology will see an ever-increasing role of space-based services for security applications, for example in the area of air and maritime transport via an improved system of traffic management or in supporting development in less advanced countries where space systems provide low-cost access to communication and to resource management. The role of space systems in overall security and defence will continue to increase, if only to provide an added degree of transparency and access to information.

However, space systems in low-Earth orbit are relatively fragile and will need to be protected if the world community wants to benefit from the sustainable use of outer space. Discussions on recommended "best practices" in outer space need to be initiated and conducted at the international level to ensure that space used for security is not jeopardised by the lack of security in outer space.

[215] United Nations General Assembly. Future Role and Activities of the Committee on the Peaceful Uses of Outer Space. Working Paper submitted by the Chairman, UN Doc. A/AC.105/L.268, 10 May 2007.

[216] Treaty on the Prevention of the Placement of Weapons in Outer Space, the Threat or Use of Force against Outer Space Objects (PPWT).

CHAPTER 8

CONCLUSIONS AND RECOMMENDATIONS

8.1 Conclusions and recommendations

Kai-Uwe Schrogl, Charlotte Mathieu & Agnieszka Lukaszczyk

The first and most visible conclusion to be drawn from the articles combined in this volume is that space applications provide almost endless opportunities for dealing with problems related to sustainability. Space applications have become important, if not indispensable tools for supporting solutions to natural and human-made problems facing global society. Threats and risks can be projected and sustainability can be promoted through the use of satellite telecommunications, navigation and Earth observation. The articles show a tremendous wealth in single applications as well as perspectives for the future. The European Space Policy Institute has recently prepared a study entitled "Case for Space"[217] which informs decision-makers about hundreds of space application opportunities. The study provides both striking examples of space applications and analyses of their foreseen political impact.

The second conclusion therefore tells about the character of space applications as an issue area of its own in international politics. This issue area is constituted by indigenous aspects of satellite applications, e.g. matters of technology development, trans-boundary cooperation and regulatory aspects. It is also constituted by elements derived from other policy areas, thus making space applications an intersection of those. The approach of this volume has identified the following six areas as the sources of space applications' setup: knowledge, environment, resources, energy, mobility and security. Figure 1 visualises this composition.

The guiding questions throughout this volume have therefore been the identification of the policy aspects of space applications related to each of the six constituting policy areas, and also of the overlapping issues which concern all policy areas. As to the first question, the method used was to have the issue areas introduced by specialists in the broader fields of "knowledge", "environment", "resources", "energy", "mobility" and "security". These more general papers then served as an input for the dialogue with space experts to examine and illustrate in detail how space contributes to sustainability and how this interaction can be made even more efficient and focused. The results in this respect have been varied, since space applications can play important roles in each of the fields but their potential has hardly been fully exploited. This is true for the issue area of energy, for example, where large opportunities are developed only slowly. This is quite different in the issue-areas of resources or environment, where satellite images are not only used

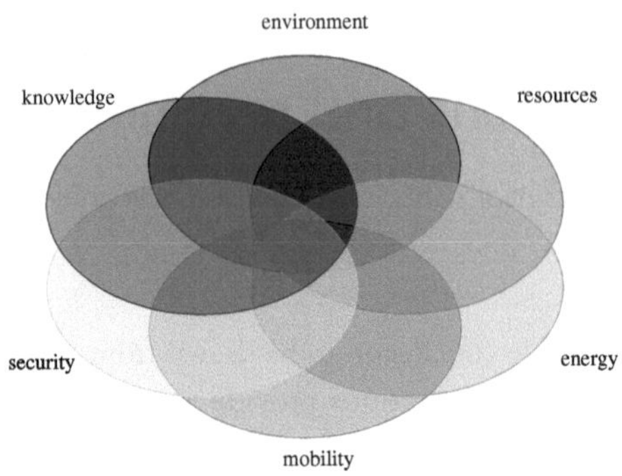

Fig. 1. *Space applications as an intersection of different policy areas.*

for detection and observation but also for verification means as a basis for treaty implementations.

The needs for action with a special focus on Europe, as identified for the six issue-areas, are the following:

(1) *Knowledge:* Europe has to raise its research performance and visibility, and overcome the fragmentation of research and funding activities. There is a need for outreach and education to bring space science to schools as early as possible. Researchers need competitive career opportunities. Trans-disciplinary research should be supported and European excellence in frontier research should be promoted.

(2) *Environment:* a better organisation of the demand and supply system is required specifically by combining satellite telecommunications, navigation and Earth observation. The value of integrated space services has to be demonstrated more precisely to users, taking advantage of the model for the development of Earth observation services within the GMES initiative. Users have to be identified more exactly. The space-based assets and systems in place for many years have not yet been used on a truly operational basis to the best of their abilities. An annual European report on the status and utilisation of space assets could be useful. Eumetsat should be empowered to be developed further as a European operational entity.

(3) *Resources:* the task is to turn information into knowledge, which has to be put into an agenda which ultimately has to be transformed into action. Clarification is needed as to who is paying for the services, who is providing the services and who is implementing the corresponding policies. Public customers, being the primary consumers of information services, must be enabled to implement corresponding budgets within their administrations. There is a need for enhancing practical

collaboration and cooperation between the various national and international agencies. Public opinion should be able to support policies for solving long-term problems. Capacity building in the use of space-derived information, like remote sensing data, is a priority for developing countries that have to build partnerships for management purposes.

(4) *Energy:* in the energy industry, in particular smaller companies, the use of space for operational services has to be increased through creating a stronger awareness of the potential benefits. At the same time, the few existing satellite systems being useful in this field have to be sustained through back-up and follow-on systems.

(5) *Mobility:* Galileo will be helpful but many issues will remain: the system will offer an unprecedented reliability in navigation services; however, privacy issues may arise in the future and thus will have to be tackled. Galileo will ensure sustainability in mobility, but will also be used in the future to restrict or limit mobility, which has to be communicated to the users.

(6) *Security:* for security on Earth, access to data should be made easier in particular regarding the broader aspects of security. For security in space, rules of the road in the form of Space Traffic Management for secure and peaceful space activities have to be developed and agreed upon at the international level.

Global development and sustainability on Earth is challenged, at least until 2050, by strong population growth, more severe environmental strains, growing mobility, and dwindling energy resources. All these factors will lead to serious consequences for humankind. Inadequate agricultural resources, water supply and non-renewable energy sources, epidemics, climate change, and natural disasters will further impact heavily on human life.

Apart from the identified needs for action in the six issue areas, some general challenges for space applications can been identified and laid down as recommendations:

- The utilisation of space has to be better organised by systematically identifying and developing users and defining operational services not yet covered by existing initiatives.
- Experimental space systems and programmes should be more easily transitioned into operational use.
- The space environment must also be sustained if it is to continue to be of benefit for future tasks.
- Europe has to make all efforts in broadening its knowledge base, in organising its own assets accordingly (with a particular role for meteorological satellites and

GMES) and to support developing regions and countries in their attempt to increase their capacities.

- Data and information sharing between states, international organisations and private actors is necessary for solving and preventing problems that affect all humankind. Space-based information is important for security on Earth and in a broader way for sustainability on Earth. It should be made freely available as a "universal service".

Coordination between different issue areas facing problems that are related to each other has to be enacted. This includes a structured dialogue with the space sector in order to find concrete and practical solutions that space as a tool can provide.

[217] Doldirina, Catherine. Case for Space: Space Applications Meeting Societal Needs. ESPI Report 7. Vienna: European Space Policy Institute, October 2007. http://www.espi.or.at/images/stories/dokumente/studies/report_case_for_space.pdf

About the authors

Jacques Blamont after having completed his thesis under A. Kastler at the University of Paris, became a professor at the University where he taught Physics from 1957 to 1996. He was one of the few who were given the task of creating the French Space Agency CNES in 1962 as Scientific and Technical Director of the fledging organisation. In 1972, he became Chief Scientist and Adviser to the Director General of CNES. Jacques Blamont is an experimental physicist who turned from atomic spectroscopy to space science when, in 1959, he directed the first launches of the French *Veronique* rockets that were the starting point of space science research in Europe. He decided early on to pursue the career of an active scientist. Among his early discoveries are the conservation of phase in the multiple scattering of resonance light in atomic sublevels, the discovery of the turbo pause in the Earth's atmosphere, and the first measurement of Einstein red shift on the Sun. In the 1960s, Jacques Blamont was involved in atmospheric research, being a prime investigator on many U.S. and Soviet spacecrafts. After his discovery in the 1970s and 1980s, he became more involved in planetary research, being in that period a member of the Science Steering Groups on the NASA missions Grand Tour (renamed Voyager), Pioneer-Venus, and of the URSS mission Vega to Venus and Halley's comet, a co-investigator on Voyager, a prime investigator on Pioneer-Venus, Vega and Phobos. Starting with the Vega mission and the various missions to Halley's Comet associated with it, he promoted cooperative programmes that brought rapprochement between the two space super powers. He acted in a similar manner in organising the joint US–USSR operations linked with the Society Union's Phobos mission. For the French balloon which was programmed to be flown in the Martian atmosphere by the Society mission in 1994, he developed on the one hand the telecom relay which has been placed on Mars Observer and Mars Global Surveyor, and on the other hand the image compression azics which were used in the Clementine mission for obtaining the first complete digital multiwavelengths map of the Moon.

Gérard Brachet holds an Engineering degree from the Ecole Nationale Supérieure d'Aéronautique (1967) as well as a Master of Sciences in Aeronautics and Astronautics from the University of Washington (1968). He began his professional career in 1970 at the Centre National d'Etudes Spatiales (the French Space Agency) where he was successively Head of the Orbit Determination and Spacecraft Dynamics Department from 1972 to 1974 in the Brétigny space centre. He then moved to CNES headquarters in Paris and took the position of Head of the Scientific Programmes Division from 1975 to 1978 and Head of the Application Programmes Division from 1979 to 1982. In 1979, he was the French signatory for the international COSPAS- SARSAT programme for satellite-based search and rescue, and subsequently the chairman of the steering committee for the same programme. He was also Chairman of the European Space Agency's Programme Board for Earth Observation from 1979 to 1981. In this capacity, he helped define the ERS-1 and ERS-2 Earth observation programmes. In 1982, Gérard Brachet left CNES and took the position of Chairman and Chief Executive Officer of SPOT IMAGE and remained in this function until 1994. From November 1996 to July 1997, Gérard Brachet was Chairman of the international "Committee on Earth Observation Satellites" (CEOS). He became Director General of CNES in July 1997 and held this position until September 2002. From October 2002 to April 2003, Gérard Brachet was advising the French Minister for Research and New Technologies on the Galileo programme. Since May 2003, he has been advising major aerospace companies as well as governments and non-governmental organisations in Europe, in particular the European Commission, on space policy and space applications. In June 2004, Gérard Brachet was selected to become Chairman of the United Nations Committee for the Peaceful Uses of Outer Space (UNCOPUOS) for the period 2006–2008, a chairmanship that he took officially at the opening of the Committee's 49th session in June 2006. Gérard Brachet is Officier de l'Ordre National du Mérite and Officier de la Légion d'Honneur.

Mario Caporale has been with the Italian Space Agency (ASI) since 2000 and is currently Head of Satellite Navigation. He represents ASI nationally and internationally on matters pertaining to Galileo. He is a delegate to the European Space Agency's Program Board on Satellite Navigation, an advisor to the EU Galileo Supervisory Board, and a member of the Galileo Signal Task Force. He is also a member of the EU/USA National Security Compatibility Compliance Group and the European Radio Navigation Plan Group. Previously, he co-chaired the GNSS Action Team of the United Nations Committee on the Peaceful Uses of Outer Space (UNCOPUOS). Early in Galileo's development and prior to joining the Italian Space Agency in 2000, Mario Caporale worked for aerospace companies, first as an avionics manager and then for 10 years as an account manager for civil defence for Telespazio SpA in Rome.

Kashif Din graduated in Electronic Engineering (Dipl.-Ing.) with a focus on computer science and telecommunications in 2003. He also obtained a M.Sc. in Information Systems Management with a focus on radio frequency identification and business process management in 2007. During his time at the university, he gained first professional experience while working for an EU-funded satellite research project dealing with low-Earth orbit (LEO) satellite communication. Currently, he is working as a project manager and researcher at Arsenal Research in the business unit of Transport Technologies. His main area of work is related to classification methods with multivariate analysis and artificial intelligence as well as information processing in cooperative vehicle infrastructure systems. At present, he is concerned with developing methods for real-time situational awareness in traffic systems based on driver and vehicle modelling. He is involved in national and international research projects (EU-FP6, FP7) aiming to enhance traffic safety and traffic efficiency with telematic solutions.

Erwin Duhamel works for ESA as the Head of the Security Strategy and Partnership Development Office in the Director General's Policy Office. He has been responsible for the development and implementation of the security strategy in the Agency since September 2007. Before that, he was the Head of the Belgian Defence Space Office. He is currently Chair of the NATO 'SSTAG' (Space Science and Technology Advisory Group) and the 'Space Systems Related Technology Group' of the EDA (European Defence Agency).

Heinz Gärtner has been professor at the Austrian Institute for International Affairs (OIIP) in Vienna since 1979. He had visiting fellowships at the Institute for East-West Security Studies in New York (1987–1989), visiting professorships at the University of Erlangen, Germany (1994/1995), the University of Vienna (1999–2001), St. Hugh's College, Oxford (1992), the University of British Columbia (1993) and the Institute for International Relations, Vancouver. He was the Austrian Chair at Stanford University, CA (2001–2002). Heinz Gärtner also participated in the International Visitors Program at the University of New Haven (with participation in programmes at Yale and Hartford) in February/ March 2005. He was a Fulbright professor at the Center for International Security and Cooperation (CISAC) at Stanford University in summer/fall 2008. He has been a regular speaker and has given presentations at universities such as Harvard, Stanford, Princeton, Columbia, Oxford and many other U.S., European, and Asian Universities and research institutes. In addition, Heinz Gärtner has been an analyst in newspapers, on radio and television and has published widely on European security, NATO, and transatlantic relations.

Menghestab Haile is currently Food Security Advisor at the United Nations World Food Programme (WFP). He coordinates the integration of advanced GIS and remote sensing applications for food security risk analysis and monitoring to enable WFP and the humanitarian community to deliver timely responses. He is also responsible for integrating climate and weather-related information into food security risk analysis and early warning. Menghestab Haile joined WFP in 1999 in the Republic of Sudan, where he worked as a Vulnerability Analysis and Mapping Officer until he moved to Rome at the end of 2002. Before joining WFP, he worked for Scot Conseil, a French company, as a regional remote sensing officer covering the IGAD countries based in Kenya. He received his Ph.D. in Meteorology from the University of Reading (UK) in 1994 and worked at the University until his return to Ethiopia in 1996 to join the Disaster Prevention and Preparedness Commission (DPPC) as a national food security early warning advisor. During that time, he contributed to the modernisation of the Ethiopian Early Warning System, including the introduction of computers and email facilities making the DPPC one of the first internet users in Ethiopia. He started his career as an assistant lecturer in Physics at the Water Technology Institute in Arba Minch (Ethiopia) in 1986, following his graduation from Asmara University (Eritrea) with a B.Sc. in Physics.

Rupert Haydn studied Geology at the University of Munich (Dipl.-Geol.) and Clausthal-Zellerfeld (Dr. rer. nat). He started his professional career as a research fellow and later as a staff member at the Geological Institute of the University of Munich where he concentrated his activities on the development of digital image processing methods, applied to multispectral satellite data. He was a principal investigator of the Heat Capacity Mapping Mission (NASA-HCMM). Having been active in the field of remote sensing since 1972, he founded GAF AG in 1985, which developed into a leading international EO and geo-information service company. He served as CEO of GAF AG since its foundation and retired from this position in 2007. Rupert Haydn has been an advisor to ESA and other international organisations and is currently a member of the supervisory board of GAF AG and Managing Director of Euromap, a subsidiary company of GAF AG.

Andreas Laake is currently a principal research scientist at WesternGeco, which is part of the leading international oilfield services corporation Schlumberger Ltd. He is based in Cairo and London. He received a Ph.D. and a Masters in Physics, Geophysics and Geology from the University of Kiel in Germany and has worked as a research scientist at the Max-Planck-Institute for Hydrodynamic Research in Göttingen, Germany. During his 20 years in the hydrocarbon industry, he worked in worldwide onshore and offshore exploration research and started the development of Schlumberger's proprietary land seismic acquisition system. For 10 years and in cooperation with ESA, he has been developing a range of methods for characterising the Earth's surface and its geology from remote sensing data, the results of which are integrated with subsurface 3D geophysical data.

Jérôme Lafeuille is Chief of the Space-based Observing System Division, WMO Space Programme Office of the World Meteorological Organization (WMO),

Geneva, Switzerland. He graduated from the Ecole Polytechnique (Paris, France) in 1975 and the Ecole Nationale de la Météorologie in 1977. He served at Météo-France from 1977 to 1992, where he was successively Director of the Snow Research Centre, Head of Strategic Planning, and Head of the Space Affairs Office. In 1993, he joined the European Organisation for the Exploitation of Meteorological Satellites (EUMETSAT) as Head of International Affairs, until he became Director of the Centre for Satellite Meteorology of Météo-France in Lannion in 1998. He joined the World Meteorological Organization in 2005.

Teresa Lago graduated from the University of Porto, Portugal in 1971 (with a B.Sc. in Mathematics; Licenciatura in Surveying Engineering). She obtained a Ph.D. in Astronomy from Sussex University, UK in 1979, and has been a full professor at the School of Sciences of the University of Porto since 1989. Teresa Lago is the founder of the Centre for Astrophysics (1987) and has been its director for 19 years. Her most relevant activities include: research in Astrophysics; teaching in Astronomy; and the promotion of science and scientific culture. She was coordinator of the Scientific Council for "Earth & Space Sciences", National Research Council (2002–2005) and a member of EC Panels DG XII, FP4, and FP5. She is a science advisor to various international committees and boards. Her memberships include, inter alia: Member of the Space Science Advisory Committee, ESA (2001–2004), Member of the ESO Council (2001–), and President of "Porto 2001 – European Capital of Culture", a 250 million euro multidisciplinary cultural project (1999–2002). Teresa Lago received the Henri Chrétien Award of the American Astronomical Society in 1985. She has also been Associate of the Royal Astronomical Society, UK, since 1990, Member of the Academiae Europeae since 1992, and a founding member of the Scientific Council of the European Research Council.

Agnieszka Lukaszczyk has been the Executive Officer of the Space Generation Advisory Council (SGAC) since September 2006. Starting from May 2008, she has also been working as a space policy consultant at the Secure World Foundation (SWF). In addition, during the period of September 2006 to June 2008, Agnieszka Lukaszczyk worked at the European Space Policy Institute. She holds a Masters degree in International Politics from the American University's School of International Service, and a Bachelor degree in Political Science from the University of Tennessee. She also studied at the Univeriste Catholique de Louvain in Brussels, Belgium; the Jagiellonian Univeristy in Krakow, Poland, and the World Trade Institute in Berne, Switzerland. She interned at the Political Section of the Polish Embassy in Washington, D.C., the American Electronics Association in Brussels, the European Department of the Polish Senate in Warsaw, and the Warsaw Business Journal.

Charlotte Mathieu has been a research fellow at the European Space Policy Institute (ESPI) in Vienna since October 2006. Before joining ESPI, she was a research assistant at the Massachusetts Institute of Technology (MIT) where she conducted research on "fractionated" satellite architectures for the Defense Advanced Research Projects Agency (DARPA) for 2 years. In 2003–2004, she was a Young Graduate Trainee in the Cost Engineering Division at the European Space Research and Technology Centre (ESTEC). She interned at the CNES Washington Office, at EADS Launchers and Snecma. She holds a double Masters degree in Aerospace Engineering from the Ecole Centrale Paris, France, and the Royal Institute of Technology (KTH), Sweden, as well as a second Masters degree in Technology and Policy from the Massachusetts Institute of Technology (MIT), USA. She also holds a degree in Economics from the Université Pierre Mendès, France.

Michael Meyer zu Hörste studied Mechanical Engineering at the Technical University of Braunschweig. From 1995 to 2001, he worked as a researcher at the Institute of Control and Automation Engineering of the Technical University of Braunschweig. In 2004, he received his Ph.D., his thesis being entitled "Methodical Analysis and Generic Modelling of Railway Operations and Train Control Systems". Since 2001, he has been a researcher at the Institute of Transportation Systems of the German Aerospace Centre in Braunschweig. Michael Meyer zu Hörste was a member of the European Rail Research Advisory Council (ERRAC) in the working team "Interoperability" as well as the Eurnex Pole 8 "Infrastructure and Signalling".

Pierre Potin is GMES Programme Coordinator at the European Space Agency, within the GMES Space Office. The GMES Space Office is the focal point for GMES within ESA and is in charge of the coordination of GMES activities within the Agency and with external organisations, in particular the European Commission. This includes the preparation of programmatic and financial documents and procedures as well as the coordination of internal activities related to the development of the GMES space ("Sentinels") and ground segments. Pierre Potin has a Masters degree in Space Systems from the Ecole Nationale Supérieure des Télécommunications, France. He has 15 years of working experience in the domain of satellite Earth observation. He initially worked on meteorological satellite programmes at a major European satellite manufacturer. Pierre Potin spent most of his career at ESA/ESRIN in Italy, with strong involvement in the Envisat mission and more recently in the GMES programme.

Kai-Uwe Schrogl is Director of the European Space Policy Institute (ESPI) in Vienna since September 2007. Before that, he was Head of the Corporate

Development and External Relations Department of the German Aerospace Center (DLR). In his previous career, he worked with the German Ministry for Post and Telecommunications and the German Space Agency (DARA). He has been a delegate to numerous international forums and recently served as the chairman of various European and global committees (ESA International Relations Committee, UNCOPUOS working groups). Kai-Uwe Schrogl has published seven books and more than 100 articles, reports and papers in the fields of space policy and space law as well as telecommunications policy. He is a member of the board of directors of the International Institute of Space Law, Member of the International Academy of Astronautics (chairing its Commission on Policy, Economics and Law) and the Russian Academy for Cosmonautics, as well as a member of the editorial boards of international journals in the field of space policy and space law (Acta Astronautica, Space Policy, Zeitschrift für Luft- und Weltraumrecht, Studies in Space Law/Nijhoff). He holds a doctorate degree in Political Science, lectures in International Relations at Tübingen University, Germany (as a Honorarprofessor) and has been a regular guest lecturer at the International Space University and the Summer Courses of the European Centre for Space Law, among others.

Wolfgang Seboldt studied Mathematics (Dipl.-Math.) and Physics (Dr. rer. nat) at the Universities of Bonn and Bochum. In 1975, he was an assistant professor at the Ruhr University Bochum (Institute of Theoretical Physics). Since 1985, he has been working for several divisions (Future Programmes, Space Systems Analysis) and institutes (Head of the Mission Architecture and Advanced Technologies Section) of the German Aerospace Centre (DLR) in Cologne. Wolfgang Seboldt has also published articles in scientific journals and books about plasma-astrophysics and planetary research, and later on also about innovative space technologies (e.g. propulsion, solar sails, in-situ resources utilisation and solar power from space). He has supervised many Masters and Ph.D. students and has been a study lead for several national and ESA studies. Wolfgang Seboldt has been a consultant to ESA for finding future strategies in space exploration and human space flight. He has also been a visiting lecturer at the International Space University in Strasbourg for many years.

Mahendra Shah is a senior scientist and Coordinator of United Nations Relations at the International Institute for Applied Systems Analysis (IIASA). His current work is concerned with sustainable development and integrated social–environmental–economic policy analysis, human capital, international negotiations, and the Millennium Development Goals. From 1997 to 2000, Mahendra Shah served as Executive Secretary to the CGIAR System Review at the World Bank and co-authored "Food in the 21st Century – From Science to Sustainable Agriculture" with Maurice Strong. He was Special Advisor to the Secretary General of UNCED in 1991–1992 and prepared the Earth Summit report "The Global Partnership for Environment and Development – A Guide to Agenda 21". In 1988–1989, he served in the UN Office of the Coordinator for Afghanistan as Director of Information, and in 1984–1986, he was Director of the Monitoring and Evaluation Division, UN Office for Emergency Operations in Africa. From 1977 to 1983, he was a senior scientist with the Food and Agriculture Program at IIASA and co-authored the UNFPA/FAO/IIASA report "Land Resources for Populations of the Future". He returned to the Institute in February 2001. Mahendra Shah received his Ph.D. from the University of Cambridge in 1971 and started his career at the University of Nairobi (1971–1977) and the Kenya Ministry of Economic Planning (1974–1976).

Mark Williamson is an independent Space Technology Consultant and technical author with some 27 years experience in the space industry. He provides consultancy services to the space insurance sector and the wider space community.He is the author of six books, including a student text, a history of spacecraft technology, the "Cambridge Dictionary of Space Technology" and "Space: The Fragile Frontier". He has also served as the editor of three international space industry magazines, written some 400 articles on space technology, and published papers on space law, space ethics and the sustainability of space exploration and development. As part of a commitment to space education, he has lectured at all levels from lay to professional and developed team-building workshops for schools and universities.He is a Chartered Engineer, Chartered Physicist and an External Examiner for Salford University (UK).

Jean-Claude Worms is Head of the Space Sciences Unit of the European Science Foundation, managing the European Space Sciences Committee and all space programmes of the ESF. He holds a Ph.D. in Physics from the University of Paris 6 and was an assistant professor in Physics and Astronomy (Paris 6 and Versailles) from 1989 to 1992. He is an associate researcher to the Laboratoire de Physique et Chimie de l'Atmosphère, the Laboratoire des Sciences de l'Image et de la Télédétection, the Laboratoire des Systèmes Photoniques, and Service d'Aéronomie. He has been the main scientific organiser and editor of solar system sessions in COSPAR Scientific Assemblies since 1998 and is a member of the editorial board of the International Journal on Nanotechnologies. In 1994, he was also a consultant for Dassault on the state-of-the-art French civilian research in infrared and SAR imaging. Jean-Claude Worms is involved in ESA and EC high-level science advisory structures, and has had observer status at ESA's Ministerial Conferences since 1999. As a result of the unique structure of the ESSC which reflects the whole variety of space-related disciplines, he is dealing with strategic planning, programme evaluation and reviewing, and intelligence monitoring in every sector of space sciences, including space policy and GMES.

Galina Zlatareva is an energy consultant in Brussels. Previously, she worked at the European Parliament, DG IPOL, Policy Department for Economic and Scientific Policies, being involved in the negotiations on the third legislative package for the liberalisation of the electricity and gas market of the EU. Galina Zlatareva holds a Masters degree in Peace Research and Conflict Management from the Institute of Political Sciences in Tübingen, Germany, as well as a Masters in European Studies and a Bachelor in International Relations. She has done a traineeship at the Ministry of Foreign Affairs in Bulgaria and was an assistant-trainee at MEP's Strasbourg office. Galina Zlatareva has published articles in the field of transatlantic economic relations and on the legal implications of the EU and U.S. policy towards supplier countries.

List of acronyms

A

AC: Alternating Current
ACARE: Advisory Council for Aeronautics Research in Europe
ADAS: Advanced Driver Assistance Systems
AES: Aeronautical Earth Station
AEZ: Agro-Ecological Zone
AIS: Automatic Identification System
AMESD: African Monitoring of Environment for Sustainable Development
AMIS: African Union Mission in Sudan
ARTIST: Architettura Telematica Italiana per il Sistema dei Trasporti (Italian Telematic Architecture for Transport Systems)
ASAR: Advanced Synthetic Aperture Radar
ASI: Agenzia Spaziale Italiana (Italian Space Agency)
A-SMGCS: Advanced Surface Movement Guidance and Control System
ATC: Ancillary Terrestrial Component
ATS: Applications Technology Satellite
AU: African Union

B

BLS: Basic Linked System
BMELV: Bundesministerium für Ernährung, Landwirtschaft und Verbraucherschutz (Federal Ministry of Consumer Protection, Food and Agriculture, Germany)
BNSC: British National Space Centre

C

C2: Command and Control
CD: Conference on Disarmament
CEOS: Committee on Earth Observation Satellites
CFSP: Common Foreign and Security Policy
CGMS: Coordination Group for Meteorological Satellites
CIA: Central Intelligence Agency
CIMIC: Civil and Military Cooperation
ClimDev Africa: Climate for Development in Africa Programme
CME: Coronal Mass Ejection

CNES: Centre National d'Etudes Spatiales (French Space Agency)
CNRS: Centre National de la Recherche Scientifique (National Center for Scientific Research, France)
CNSA: China National Space Administration
CONAE: Comisión Nacional de Actividades Espaciales (Argentine Space Agency)
COPUOS: Committee on the Peaceful Uses of Outer Space
COSPAR: Committee on Space Research
CRM: Climate Risk Management
CRU: Climate Research Unit
CSA: Canadian Space Agency
CSD: Corporate Sustainable Development

D

DC: Direct Current
DEM: Digital Evaluation Map
DEM: Digital Elevation Model
DG: Dangerous Goods
DGT: Dangerous Goods Transportation
DLR: Deutsches Zentrum für Luft- und Raumfahrt (German Aerospace Center)
DSMW: Digital Soil Map of the World

E

EADRCC: Euro-Atlantic Disaster Response Coordination Centre
EAGGF: European Agricultural Guidance and Guarantee Fund
EAT: Energy Amortization Time
EC: European Commission
ECV: Essential Climate Variables
EDA: European Defence Agency
EGNOS: European Geostationary Navigation Overlay Service
EO: Earth Observation
EQHHPP: Euro-Québec Hydro-Hydrogen Pilot Project
ERC: European Research Council
ERRAC: European Rail Research Advisory Council
ERTMS: European Rail Transport Management System
ERTRAC: European Road Transport Research Advisory Council
ESA: European Space Agency
ESDP: European Security and Defence Policy
ESPI: European Space Policy Institute
ESRO: European Space Research Organisation

ESS: European Security Strategy
ESTP: European Space Technology Platform
ETa: Actual Evapo-Transpiration
ETCS: European Train Control System
ETP: European Technology Platform
EU: European Union
EUMETSAT: European Organisation for the Exploitation of Meteorological Satellites

F

FAO: Food and Agriculture Organization
FAPAR: Fraction of Absorbed Photosynthetically Active Radiation
FCD: Floating Car Data
FC Evaluator: Floating Car Evaluator
FEWSNET: Famine Early Warning System
FLD: Forest & Landscape Denmark
FRC: Functional Road Class

G

GCMs: General Circulation Models
GCOS: Global Climate Observing System
GDP: Gross Domestic Product
GENESI-DR: Ground European Network for Earth Science Interoperations – Digital Repositories
GEO: Geostationary Orbit
GEO: Group on Earth Observation
GEOSS: Global Earth Observation System of Systems
GES: Ground Earth Station
GHG: Greenhouse Gases
GIEWS: Global Information and Early Warning System (on Food and Agriculture)
GIS: Geographic Information System
GLEN: Global Energy Network
GLONASS: Global Orbiting Navigation Satellite System
GMES: Global Monitoring for Environment and Security (now: Kopernikus)
GMFS: The Global Monitoring for Food Security
GNSS: Global Navigation Satellite Systems
GPS: Global Positioning System
GSA: GNSS Supervisory Authority

GSE: GMES Service Elements
GSE FM: GSE Forest Monitoring

H

HCR: High Commission for Refugees
HGT: Hazardous Goods Transport
HLLV: Heavy Lift Launch Vehicles
HV: High-Voltage
HVAC: High-Voltage Alternating Current
HVDC: High-Voltage Direct Current

I

IACS: Integrated Administration and Control System
IADC: Inter-Agency Space Debris Coordination Committee
IAEA: International Atomic Energy Agency
ICPAC: IGAD Climate Prediction and Analysis Center
ICT: Information and Communication Technology
IEA: International Energy Agency
IIASA: International Institute for Applied Systems Analysis
IM: Image Mode
IMO: International Maritime Organization
InSAR: Interferometric Synthetic Aperture Radar
IOV: In-Orbit Operational Validation
IPCC: Intergovernmental Panel on Climate Change
ISRO: Indian Space Research Organization
ITS: Intelligent Transport Systems
ITU: International Telecommunication Union

J

JAXA: Japan Aerospace Exploration Agency

K

KP: Kyoto Protocol

L

LAI: Leaf Area Index
LBS: Location-Based Services
LEO: Low Earth Orbit
LPS: Lunar Power Stations
LUT: Land Utilization Type

M

MARS: Monitoring Agriculture with Remote Sensing
MASLI: Indonesian Society for Natural Resources and Environmental Accounting
MATT: Ministero dell'Ambiente e della Tutela e del Mare (Ministry for the Environment and Territory, Italy)
MDA: Massive Destruction Armaments
MDGs: Millennium Development Goals
MEDUP: Mediterranean Upgrade Programme
MEO: Medium Earth Orbit
MFF: Mediterranean Free Flight
MOC: Meridional Overturning Circulation
MRD: Mission Requirement Document
MSAS: Multifunctional Satellite Augmentation System
MTB: Mediterranean Test Bed
MTG: Meteosat Third Generation
MUSIS: Multinational Space-Based Imaging System

N

NAC: North Atlantic Council
NASA: National Aeronautics and Space Administration
NATO: North Atlantic Treaty Organization
NEO: Near-Earth Object
NGOs: Non-Governmental Organisations
NOA: National Observatory of Athens
NOAA: National Oceanic and Atmospheric Administration
NPS: Nuclear Power Sources
NRF: NATO Response Force

O

OBUs: On-Board Units
OCHA: Office of Coordination of Humanitarian Affairs
OPEC: Organization of the Petroleum Exporting Countries
OSCE: Organization for Security and Co-operation in Europe
OST: Outer Space Treaty
OT: On-Board Terminal

P

PAROS: Prevention of an Arms Race in Outer Space
PDA: Personal Digital Assistant

PNT: Positioning-Navigation-Timing
Post-EPS: Post EUMETSAT Polar System
P-RNAV: Precision-Radio Navigation
PRS: Power Relay Satellites
PV: Photovoltaic

R

RCMRD: Regional Centre for Mapping Resources for Development
R&D: Research and Development
RTK: Real Time Kinematics

S

SADC-REWU: SADC Regional Early Warning System for Food Security
SAR: Synthetic Aperture Radar
SBA: Societal Benefit Areas
SBAS: Space Based Augmentation Systems
SBSP: Space Based Solar Power
SCTM: Space Solar Power Concepts & Technology Maturation
SDLS: Satellite Data Link System
SEGS: Solar Energy Generating Systems
SERT: Space Solar Power Exploratory Research and Technology
SG: Space Guard
SHAFT: Standardized Hazardous Goods Transport Alerting Field Trail
SPS: Solar Power Satellites
SRA: Strategic Research Agenda
SRES: Special Report on Emissions Scenarios
SRRA: Strategic Rail Research Agenda
SSA: Space Situational Awareness
SSP: Space Solar Power
ST: Solar Thermal
STENAV: Satellite Test Bed for Navigation and Communications
STF: Space Task Force
SWIR: Short-Wave Infra-Red

T

TCBM: Transparency and Confidence-Building Measure
TFSC: Thin Film Solar Cells
TW: Terrawatt

U

UAV: Unmanned Aircraft Vehicle
UCINA: Unione Nazionale Cantieri e Industrie Nautiche e Affini (Italian Marine Industry Association)
UN: United Nations
UNCOPUOS: United Nations Committee on the Peaceful Uses of Outer Space
UNFCC: UN Framework Convention on Climate Change
UNGA: United Nations General Assembly
UNOOSA: United Nations Office for Outer Space Affairs
U.S.: United States
USA: United States of America
USGS: U.S. Geological Survey

V

VAM: Vulnerability Analysis and Mapping
VLT: Very Large Telescope
VLTI: Very Large Telescope Interferometer
VNIR: Visible and Near-Infrared
VSAT: Very Small Aperture Terminal
VTS: Vessel Traffic System

W

WAAS: Wide Area Augmentation System
WAN: Wide Area Network
WB: World Bank
WENET: World Energy Network Using Hydrogen
WFP: World Food Programme
WHO: World Health Organization
WMD: Weapons of Mass Destruction
WMO: World Meteorological Organization
WPT: Wireless Power Transmission
WRSI: Water Requirement Satisfaction Index
WSM: Wide Swath Mode

X

xFCD: Extended FCD

List of figures and tables

Figures

Chapter 2: Knowledge

Chapter 3: Environment

Chapter 4: Resources

Chapter 5: Energy

Chapter 6: Mobility

Chapter 7: Security

Chapter 8: Conclusions and Recommendations

Tables

Chapter 2: Knowledge

Chapter 3: Environment

Chapter 4: Resources

Chapter 5: Energy

Chapter 6: Mobility

Chapter 7: Security

MIX
Papier aus verantwortungsvollen Quellen
Paper from responsible sources
FSC® C105338

If you have any concerns about our products,
you can contact us on
ProductSafety@springernature.com

In case Publisher is established outside the EU,
the EU authorized representative is:
Springer Nature Customer Service Center GmbH
Europaplatz 3, 69115 Heidelberg, Germany

Printed by Libri Plureos GmbH
in Hamburg, Germany